Machine Tools

About the Author

Russell Gamblin is currently the Maintenance and Facilities Manager at Mitsubishi Power Systems, Inc. Previously, he was a Project Engineer at Mitsubishi Power Systems and Sr. Maintenance and Facilities Engineer at Siemens Energy, Inc. Within the past three years, Mr. Gamblin has been a part of two large new factories additions and relocations with combined costs of $1 billion.

About SME

SME connects all those who are passionate about making things that improve our world. As a nonprofit organization, SME has served practitioners, companies, educators, government and communities across the manufacturing spectrum for more than 80 years. Through its strategic areas of events, media, membership, training and development, and the SME Education Foundation, SME shares knowledge to advance manufacturing. At SME, we are making the future. Together.

Machine Tools
Specification, Purchase, and Installation

Russell Gamblin

New York   Chicago   San Francisco
Athens   London   Madrid
Mexico City   Milan   New Delhi
Singapore   Sydney   Toronto

Cataloging-in-Publication Data is on file with the Library of Congress.

McGraw-Hill Education books are available at special quantity discounts to use as premiums and sales promotions, or for use in corporate training programs. To contact a representative please visit the Contact Us page at www.mhprofessional.com.

Machine Tools: Specification, Purchase, and Installation

1 2 3 4 5 6 7 8 9 0   QVS/QVS   1 2 0 9 8 7 6 5 4 3

ISBN 978-0-07-181222-1
MHID 0-07-181222-9

The pages within this book were printed on acid-free paper.

Sponsoring Editor	Proofreader
Judy Bass	Ragini Pandey,
Acquisitions Coordinator	Cenveo Publisher Services
Amy Stonebraker	**Indexer**
Editorial Supervisor	Robert Swanson
David E. Fogarty	**Production Supervisor**
Project Manager	Lynn M. Messina
Raghavi Khullar,	**Composition**
Cenveo® Publisher Services	Cenveo Publisher Services
Copy Editor	**Art Director, Cover**
Patti Scott	Jeff Weeks

Contents

Preface

In this book, I offer the extensive knowledge gained from my years of purchasing, installing, and maintaining machine tools and equipment.

First, you will find a systematic approach to the specification and procurement phase. You will learn how to weed out the machine tool suppliers you don't need, and you will learn how to develop a step-by-step plan for selecting the *right* machine tool for your company.

As we move through the text, we will go deeper into the specification and build of the machine tool's foundation. You will become familiar with the cardinal rule that one should *never* build the foundation exactly to the machine tool supplier's specifications. Rather, the machine tool's foundation must fit your facility *first*. Functionality comes second. Build the foundation that fits the specific machine tool *and* the environment in which it will operate.

This book isn't about theoretical formulas associated with machine tools. Rather, it offers real-world examples of what does and does not work. It's full of stories that illustrate things you should and shouldn't do—stories that come from my own hands-on experience.

I have heard so many horror stories in my time—stories of bad machine tool purchases and the many problems stemming from poorly planned and poorly constructed foundations. It became clear to me that what was needed was a book to provide sound, time-tested knowledge to people in the machine tool industry who could truly benefit from it.

This book offers, in clear and concise terms, answers to the multitude of questions that I have had to find answers for throughout my own career. It is a user-friendly resource for the novice as well as the seasoned professional. However, it will prove most beneficial to those who are project leaders and members of the project team tasked with specifying, purchasing, and installing the new machine tool.

Writing this book has been both humbling and rewarding. It is not something I could have accomplished alone. Therefore, I would like to thank the following people for their invaluable contributions to the completion of this work:

I thank my parents, Russell and Sharon Gamblin, who taught me that working hard and staying focused will get me what I want in life, and not to concern myself with what others may think.

I would like to thank Jack Bradshaw, Bill Carriker, Troy Starr, Bruce Edwards, and all the maintenance and facilities guys at Siemens and the old Westinghouse facility in Charlotte, North Carolina. Also I would like to mention Dane Sebastian and David Mazelin from Siemens and Westinghouse who helped me understand that you must keep the purchasing department involved in the process. The purchasing department will likely keep you out of trouble after the machine tool is installed and will help you understand the terms and conditions of the purchasing contract.

I would especially like to thank Steve Bowman and Mike Mundy of Siemens—two people whom I lived with in the trenches of the daily grind of the machine tool industry, two people who rarely get the credit they deserve.

I would also like to thank Greg Martin, Sam Suttle, and Jack Daniels for taking a chance on me at Mitsubishi Power Systems in Pooler, Georgia. They allowed me to do what I thought was needed to properly install some of the largest machine tools in the world, and they backed me when most other people thought I was crazy.

Most of all, I want to thank my wonderful wife, Amy, and two sons, Russell Ray Gamblin, III (Trey), and Parker Gamblin. They are the reason I live today and why I continue to succeed in life.

Thanks also to writer Clark Byron for helping me organize my thoughts and convey them to the reader in the most accessible terms.

Russell Gamblin

CHAPTER 1
Budget Phase

1.1 Attend One of the Many Quality International Trade Shows

The International Machine Tool Show (IMTS) occurs every 2 years at the McCormick Center in Chicago, Illinois. It is the premiere machine tool trade show in the United States.

The Europe Machine Show (EMO), formally the *Exposition Mondiale de la Machine Outil* (English: Machine Tool World Exposition), is a biannual European trade show. It is held every odd-numbered year. It is at the Hanover fairground in Hanover, Germany, for two shows and the FieraMilano Exhibition Center in Milan, Italy, for one show.

The South-TEC trade show is held yearly in Greenville, South Carolina. It is a major gathering of leading companies such as manufacturing suppliers, distributors, and equipment builders, both national and international. The exhibitors come to demonstrate their latest technologies.

The value of attending at least one or two of these events cannot be overstated. Before you purchase any large machine tool, it is essential to be exposed to the newest technologies and the wealth of information provided at these events. Expect to spend at least 3 or 4 days scouting at each show. It will require at least that much time to see all that is being offered. This is necessary preparation for deciding what machine tool is the best choice for your shop's particular needs.

It is extremely helpful to talk to the machine tool suppliers exhibiting at these shows. They will help you develop an original equipment manufacturer (OEM) list that will be the starting point for your planning and purchasing process. By the time you leave the trade show, your list should include between 10 and 20 machine tool suppliers whom you can contact for information and price quotes for the machine tool you are considering.

1.2 Develop a Return on Investment Projection

Most machine shops such as yours have a financial policy that is usually referred to as a *return on investment* (ROI) policy for capital purchases, i.e., fixed assets. In other words, what kind of a profit can you expect to realize from this machine throughout its projected lifespan? The ROI is usually calculated and submitted on a computerized form with fields that need to be populated with certain vital information.

First find out your company's ROI requirements. Then complete the form as best you can at this point. There is plenty of information you will not have this early in the process. However, starting with the ROI form is the best place to begin calculating project costs and eventually reach an ROI projection. You will not yet know the cost of the machine, let alone the overall cost of the entire project. However, your initial research and information gleaned from several sources, including the trade shows, should provide some round numbers to drop into the ROI form to calculate a very rough project cost. This can be used to estimate how much money your company should plan to spend, but keep in mind that this is very preliminary. It will most definitely change as the process moves forward.

Keep in mind that when you are trying to purchase a new machine tool, it is important to detail the issues associated with the machine tool that you are trying to replace. This information will help you justify the purchase of the machine tool and can be added to the return on investment form when it is turned in to upper management. This should include some of the following items:

1. Maintenance labor costs.

2. Downtime associated with the machine tool over the past 12 months. Get an average of the 12 months for the new machine tool justification.

3. Lost earnings due to the machine tool not being available.

4. Waiting for parts. The older the machine tool, the more difficult it is to find spare parts that are available to fix broken or damaged items on the machine.

5. Reduction of spare parts in inventory. The longer a machine tool has been running at your facility, the more spare parts you will accumulate over the life of the machine.

6. Reduction in emergent breakdowns. Breakdowns always occur at the worst times. Keep in mind that most large machine shops work 24 h/7 days/wk. If a breakdown occurs on second shift at 8 p.m., most likely there won't be any technical support until the next morning and your company would have lost more than 16 hours on the machine tool.

7. Scheduling and planning will be difficult because of unknown reliability issues associated with the machine tool. It will be almost impossible to schedule your machinists and plan machining operations with a machine tool that constantly gives you problems and breaks down.

8. New machining and tooling processes. Even though you have to spend additional capital funding and operational resources on the new machine tool, the new machining and tooling processes might help justify the new machine tool.

1.3 Prepare a Payback Projection for the Project

Along with the ROI, most companies have a payback requirement for every capital purchase. Payback is time estimation, not a financial projection, but it is based on financial information. It is the estimate of how long it will take the machine's usage to repay the original purchase price of the machine. As you did with the ROI, work with your company's finance department to determine the payback requirements. This information will be essential to choosing a new machine tool and getting the approval from the company to purchase it.

1.4 Develop a Breakeven Cost for the Machine Tool

Contrary to how it may sound, the breakeven cost is not the same as payback. Remember, payback is an estimate of *time*. Breakeven is the amount of money it will take to recover the initial investment, including fixed and variable costs associated with the purchase, installation, and operation (including maintenance) of the machine tool. It is likely that your company has set parameters for this as well. Find out what they are and, as best you can with the information available to you at this point, make a projection.

Remember that all this information in these projections will change as more-exact numbers become available throughout the course of the specification process. But these projections will be needed in helping your company choose the needed machine tool for its future machining application.

1.5 Develop the First Specification of the Machine Tool

As with any mechanical thing, machine tools have a long list of specifications that describe their many characteristics and capabilities. That's *not* what we're talking about here when we speak of a specification. In this case, the term *specification* refers to the form that details all the specifics and performance capabilities of the machine tool to be purchased; the numerous and complex details of the installation; the training necessary for the proper, safe, and profitable operation of the machine tool; and everything else needed to make a collective, wise buying decision. The machine tool specification (or just *specification* from here forward) will detail every standard feature on the machine tool as well as available options that can be added. The specification will also outline the terms and conditions of the project.

Some call this document the *purchase specification* while others prefer the term *machine tool specification*. It's the same document. As with the other projections listed in the previous segments, the specification will become increasingly refined as new and more precise information comes in prior to the purchase of the machine tool. See Table 1.1 for an example of the first-stage machine tool specification.

The specification is not a projection, but rather a plan that will eventually evolve into *the* plan. It will contain, among many other things, the actual specifications of the machine tool itself. For example, if a special foundation is required for the machine tool (which is usually the case), then a separate specification has to be prepared for the foundation.

Item	Minimum Requirements	Notes
X-axis travel (longitudinal)	120 in.	
Y-axis travel (vertical)	96 in.	
Z-axis travel (boring/quill)	36 in.	
W-axis travel	If ram is included, 24 in.	
B-axis travel	48 in. × 48 in. square table	Is the table part of the machine?
V-axis travel	36 in.	
Minimum table size if not one table	48 in. × 72 in.	
X-axis construction	Linear guideway or boxway construction	Detail which design with quote
Y-axis construction	Linear guideway or boxway construction	Detail which design with quote
Spindle horsepower	50hp	
Spindle thrust	1200 ft·lb min.	
Spindle diameter	6 in. maximum	
Spindle tool taper	CAT 50/50 taper	
Arbor tool and pull-stud details	Included drawing of arbor tool and pull-stud details	
Spindle gear ranges	Per OEM/please specify	
Spindle rpm	0 to 1500 rpm/please specify	
Coolant tank size	Min. 250 gal	
Through the spindle coolant	Yes/no	
Number of coolant connections at the front of the ram	2 minimum	Coolant at the ram face
Minimum clamping force of the retention knob/tool	5000 lb	
M19/spindle orientation included	Yes/no	
Work light included	Yes/no	
Size of the ram	12 in. × 12 in. square	If included
Size of the rotary table	Min. 48 in. × 48 in.	Or larger
Weight capacity of the table	20,000 lb minimum	

TABLE 1.1 Draft Machine Tool Specification

Item	Minimum Requirements	Notes
Locking pin for rotary table every 90°	Note if you have additional options	
B-axis construction	Linear guideway or boxway construction	
V-axis construction	Linear guideway or boxway construction	
Rapid traverse speeds		
X-axis	400 ipm	
Y-axis	400 ipm	
Z-axis	300 ipm	
B-axis	3 rpm	
V-axis	300 ipm	
Maximum cutting speed (all axes)	150 ipm	
Telescopic way covers included	Yes/no	
Control options	Fanuc, Siemens, Mitsubishi, Fagor, etc.	Please detail CNC
Tool changer	30, 60, or 90 tool holder tool changer	If tool changer included
Max. weight of tool	55 lb	If tool changer is included
Feedback system	Heidenhain or Fagor incremental tape or glass scales	
Repeatability of all axes	0.0001 in.	
Accuracy of the rotary table	±3.5 arc seconds (min.)	
Color of the machine tool	Dark gray	
Installation included	Does the cost of the machine include the installation?	Yes/no
Supply voltage—three-phase	460 VAC/3-phase/ (±10%)	
Isolation transformer included	Yes/no	

TABLE 1.1   Draft Machine Tool Specification (*Continued*)

Item	Minimum Requirements	Notes
Training		
Operator	One week of training	
Detail of operator training required		
Maintenance	One week of training	
Application	One week of training	
Spare parts	Recommended spare parts list	Include
Warranty	Minimum 12 months on machine tool and control	
Terms and conditions	20% down with purchase order	
	30% with signoff at machine tool factory	
	30% with delivery of machine tool at buyer's factory	
	20% four weeks after acceptance of machine tool at buyer's factory	
Foundation and general arrangement drawings	Include foundation and general arrangement drawings with quote	
Leveling bolts/wedges/anchor bolts included	List all leveling bolts/wedges/anchor bolts to be included	
Acceptance/alignment/laser shots included	Include geometric accuracy requirements with quote	
Machine runoff/horsepower test (min. 75%) cut		
Detail list of G and M codes for this machine tool	G and M code list required	
Manual operation mode (manual feed/spindle on)	Operator-friendly options (please detail all options)	
Detailed assembly drawings of all major machine components		

Table 1.1 Draft Machine Tool Specification (*Continued*)

Item	Minimum Requirements	Notes
Detail drawings of all reusable parts		
Cost of basic machine		
Options		
Chip conveyor		
Chip deflectors		
Right-angle head	Include specifications	
Universal head/manual adjustment	Include specifications	
Universal head/automatic adjustment	Include specifications	
Cost of machine installation	Include all expenses for machine installation	
Cost of spare parts		

TABLE 1.1   Draft Machine Tool Specification (*Continued*)

The machine tool specification provides vital information in preparing the foundation specification. If your company has engaged the services of an architectural or engineering firm to design the foundation, you may also require a customized specification that the firm will use to design the foundation.

Preparation of the specification is the responsibility of the project manager or the manufacturing engineer in your company. Everyone on the project team has a part in the process, and it will become the singular plan that everyone will use in managing and completing her or his portion of the project. It also shows, in a single document, how each part of the plan affects the others parts, helping to ensure the proper integration of all the work done by the team.

In addition to the project team members at ground level, your company's purchasing agent will reference the specification when he or she places the order for the machine tool the company has chosen. The manufacturing engineer will determine the overall travel (length of motion capabilities in making the part it is intended for) of all the axis, which impacts the space needed for the effective operation of the machine. The machine operators or *operational group* will reference the specification to know how much training they will be receiving on the new equipment. The maintenance and facilities group will consult the specification to determine if spare parts should be purchased at the time of the specification rather than waiting until later when such parts will likely be much more costly.

Now, since there is no substitute for experience, it is essential that you locate persons within your own company who have purchased (Purchasing), planned (Operations), installed (Facilities), and operated (machinists) new machine tools in the past. If you are a start-up company, it is wise to find such people outside your company who can advise you. Without such guidance, your company runs the risk of wasting millions of dollars on incorrect or unnecessary equipment purchases and improper installation and management. (See Table 1.2 for a sample Project Team Members form.)

This is my reason for writing this book: To give you and your team the information you need to create a thorough and proper plan for the purchase and installation of your new

Department	Team Member	Signoffs for the Draft Specification
Operations/ Manufacturing	Manufacturing engineer machinists	
Purchasing	Capital purchasing agent	
Facilities	Electrical facilities engineer	
	Mechanical facilities engineer	
	Plant engineer	
Maintenance	Controls engineer	
	Mechanic	
	Electrician	
Accounting	Accountant	
Engineering	Machining engineer	
Programming	NC programmer	
Quality	Quality engineer	

TABLE **1.2** Project Team Members

machine tool. The more you plan and properly manage your machine tool project from the start, the more time and money you will save at the end of the project.

The project manager has a heavy load of responsibility. First, he or she must know the specifications by heart, including the changes and updates as they occur. The project manager is the go-to person for answers and instructions. The project manager should not have to pull out the specification every time a question arises. In addition, the project manager has to be able to answer the broader questions, such as what exactly was purchased and why.

The specification serves a number of other purposes. It helps prevent financial misconduct in the purchasing of this multimillion-dollar piece of equipment. Sometimes a company repeatedly purchases only one or two types of machine tool. There is never an effort made to see what new technology is available that may meet that company's manufacturing needs more effectively than the machine tools they keep replacing with similar equipment.

The reasons for this could be many and quite innocent. Perhaps no one has been assigned to investigate the latest technologies. Maybe management feels that it's just easier to stick with the status quo, since these are the machines the operators already know how to use and are best suited for their long-standing manufacturing process. These reasons and others like them are usually signs of poor management or a lack of vision to be on the cutting edge of the industry, but they do not reflect dishonesty.

However, the reason for the repeated purchases of like equipment from the same supplier could be a sign of trouble. There may be some private arrangement between someone in the company and someone with the supplier involving a kickback of some kind. It's rare, but it happens. These are multimillion-dollar purchases and temptations can be great.

Having a team assembled and involved from the start, a team that involves everyone from managers to engineers to operational departments, is the best way to assure upper management that the company is always on the cutting edge of the industry and that no one is reaping undue benefits from the purchase of this very valuable equipment.

1.6   Develop a List of Technical Differences between the Proposed Machine Tool and the Older Machines in Your Factory

At this point, it's necessary to take stock of what is and is not working in your factory's production process. Are there machines of differing designs or technologies that are causing glitches in your production process?

Here's an example: I was working for a company that purchased a large horizontal boring machine (HBM). It was an extremely large linear guideway machine. However, most of the other HBMs were large hydrostatic boxway construction machines. These are stronger and much heavier than the linear guideway machine. Both types of HBMs had small spindles and tapers for tooling. However, there were differences in these machines that caused numerous problems. The depth cut on the new machine was much less than on the linear guideway, but the speeds and feeds on the new machine were increased. This mismatch caused many problems with programming, engineering, and maintenance. Eventually, solutions to the problems were found, but the problems and delays were costly.

This example was the result of too few experts being involved. This is why it's important to gather your most knowledgeable engineers, managers, and maintenance people together for a consensus on exactly what type of machine tool is needed before you add too many details to the specification.

After the first draft of the specification is completed, it is important for all parties, including engineering, purchasing, manufacturing, and facilities/maintenance, to sign off. This act ensures that all departments are in agreement about the purchase of the right machine tool for the company's needs.

1.7   Develop a List of Suppliers

Having as many supplier options as possible means you have a better chance of getting the best equipment at the best price and on terms that are most favorable to your company. If your company doesn't already have a substantial list of machine tool suppliers, you must make one. If there is one already, it is likely to be found in the purchasing department. Take a little time to make sure it's up to date. Your trips to the tradeshows will almost certainly provide you with new information for your list. Your focus will be on the suppliers that carry the specific type of machine tool your company is looking for.

Let's say the machine you need is an HBM. There are more than 100 manufacturers of HBMs around the world. By attending the trade shows and spending time in Internet research, you will find a host of machine tool suppliers. The more you can contact, the better your chances will be of finding just the right machine and the right terms for your project. However, keep in mind that these machines are multimillion-dollar purchases, or, in the eyes of the suppliers, multimillion-dollar sales. They don't make these kinds of sales every day. It takes a great deal of time, work, negotiation, and just good old-fashioned selling to close a deal on a machine like this one. What this means to you is that the more suppliers you contact, the more will be on your phone and at your doorstep constantly. This is not a bad thing, mind you; just be prepared for it. Keep in mind that it is an essential form of networking, and networking is vital, especially at this point, when you are beginning to spec projects and machines. We will discuss this further in Chap. 2. (See Table 1.3 for a sample machine tool supplier list.)

1.8   Rate Machine Tool Suppliers, Their Equipment, and Their Capabilities

When speaking of machine tool suppliers and machine tool manufacturers, we are speaking of the same entity. That said, there are vast differences between machine tool manufacturers. The areas of interest for you are quality, cost, name brand recognition, and reliability. Craftsmanship differs significantly from manufacturer to manufacturer. When it comes to quality, there are essentially three levels of machine tool manufacturers: high, medium, and low.

Characteristics of a high-end manufacturer are that it has more than 100 years of being in the business, thousands of machines already operating in the industry, and more than 5000 employees. Such companies are highly regarded in the industry the world over. They may also be divisions of still larger companies.

Machine Tool Supplier	Supplier Country of Origin
MAG - Giddings and Lewis	United States
MicroCut	Taiwan
Cincinnati Machine Tools	United States
Skoda Machine Tool	Czechoslovakia
Mitsubishi Machine Tool	Japan
Union Machine Tool	Germany
Ingersoll Machine Tool	United States
DorriesScharmann	Germany
Toshiba Machine Tool	Japan
TOS Machine Tool (Fermat)	Czechoslovakia
Summit Machine Tool	United States
Kitamura Machine Tool	Japan
Rebuilder of Used HBMs	**Supplier Country of Origin**
Lucas Precision (rebuilt HBM)	United States
Machine Tool Research (rebuilt HBM)	United States
Dipaolo Machine Tool (rebuilt HBM)	Canada
Total Controls Services (rebuilt HBM)	United States/ United Kingdom
Kentucky Rebuild (rebuilt HBM)	United States

TABLE 1.3   Horizontal Boring Machine Suppliers

Not surprisingly, these companies charge the highest prices for their products because they have the quality and reputation to do so. Buying a machine tool from one of them is a good risk. The chances that the tool will have problems are extremely low.

It is also possible that the high-end manufacturer with the credentials just mentioned will know your manufacturing processes even better than you do. This is so because the manufacturer sells to thousands of companies like yours including some of your competitors. It has the benefit of seeing how many companies do the same work and what the latest technologies and techniques are producing.

Medium-range manufacturers have over 50 years of experience, thousands of machines in the industry, and at least 2000 employees.

Low-end suppliers have less than 50 years in the business, fewer than 1000 machines in the field, and less than 2000 employees.

The main questions you need answered are these: How many machines of this design are currently in use in the field? Is this a new design? How many field service people does the company have in the country where the machine will be operating? Can most, if not all, replacement parts be delivered within 24 h? You might also ask each manufacturer that you're considering to fill out a questionnaire. (See Table 1.4 for a sample supplier rating form.)

Item	High-End Suppliers	Medium-End Suppliers	Low-End Suppliers
Number of machine tools in the industry	More than 5000 machine tools in the machine tool industry	1000 to 5000 machine tools in the machine tool industry	Less than 1000 machine tools in the machine tool industry
Years of experience	More than 100 years of experience in the machine tool industry	100 years to 50 years of experience in the machine tool industry	less than 50 years of experience in the industry
Cost	High Cost machine tool supplier	Medium Cost machine tool supplier	Low Cost machine tool supplier
Number of employees	More than 1000 employees	500 to 1000 employees	Less than 500 employees
Experience in the field of machine tools	Expert in the machine tool industry	Expert to Moderate in the machine tool industry	Moderate in the machine tool industry
Brand name in the machine tool industry	Known brand name in the machine tool industry	Somewhat known in the machine tool industry	Unknown to somewhat known in the machine tool industry
Risk of the machine tool not operating per the specification and process specified	Extremely low risk	Moderate risk	Moderate to high risk

TABLE **1.4** Rating the Machine Tool Suppliers

1.9   Visit Machine Tool Manufacturers

If you are in charge of the purchase of a very sophisticated and expensive machine tool, it only makes sense to visit the manufacturer or manufacturers under consideration. It is vital that you familiarize yourself with the company beyond the surface information you can gain from external sources. You must see how the company is run, what its manufacturing processes are like, and how the equipment is assembled.

In preparation for your travels, it is wise to query your machinists, supervisors, mechanics, and electricians about what questions they would ask of the manufacturer if they were going with you. It's possible you will get a question or two from your own team that will stump the manufacturer's salespeople.

The time and travel are sometimes costly but always well worth it. If for budgetary reasons, your employer does not permit travel at company expense, then your employer may want to reconsider purchasing a new machine tool at all. The importance of making these contacts personally, seeing these processes, and building these relationships for your company cannot be overstated. You need to know firsthand what you're getting for the enormous amount of money your company will be handing over to the manufacturer, to say nothing of the million or more dollars in ancillary costs you'll spend before the machine is up and running. This manufacturer will also be providing service, parts, and support for a machine you'll likely be using for the next 20 years. If this turns out to be a good purchase, the chances are very good that you'll be buying your next machine from that same manufacturer too. Now is the time to begun building relationships with a manufacturer or manufacturers that appear to be the kind your company wants to do business with.

A visit to manufacturers will also give you the opportunity to see for yourself who has the best engineering support. It will allow you to find out who the contact person will be and to meet him or her. The same goes for parts, sales, and other services. It's important to glean all this information and make all these contacts at each of the manufacturers you visit. You'll probably buy from one of them, and you'll want to know about all of them.

Transparency and candor are essential to making the right buying decision. Both the buyer (you) and the seller (them) should be completely upfront with each other. During your visit, tell the manufacturer all your concerns. Ask the right questions and get truthful answers: Were the last 10 machine tools sold installed on time? Were there any problems with any installations? What are the five most common problems they've encountered with this type of machine? If these questions make the manufacturer uncomfortable, just remember that this is the time to do that, not when you are suffering with a machine that's not meeting your needs. This is the time to find out what the people who make these machines are really made of.

1.10 Gather "Ballpark" Quotes from Each Machine Tool Manufacturer

Once you have selected the manufacturers you want to consider, the next step is to start collecting and reviewing financial information based on your first specification. Make sure every member of your team has a copy of every quote. Even if some people don't look at them, you will have done your due diligence by providing everyone with the most current information to date. Understand that it will take weeks, not days, for you and your team to perform a thorough evaluation of each manufacturer and make the necessary comparisons.

The primary question to answer from all this is which machine tool will provide the most extensive standard features for the money. For example, which one provides the most travel, the greatest spindle horsepower, and the best computer numerical control (CNC) operator control?

Let's not forget about price. Which one is offering the best price? My experience tells me that the lowest price does not always mean the wisest choice. The thing to look for is value: Are you getting the most for what you are spending? Manufacturers, more than anyone else, would agree with this approach.

1.11 Develop an Overall Budget Cost for the Machine Tool

Now is the time to develop the first complete draft budget for the machine tool that you and your team agree is the right choice for your company. Why do it this early in the process? The answer is simple: sticker shock. Most of your upper management doesn't have a good grasp of what a machine tool and the accompanying project are going to cost. It will be astronomical by anyone's standards. The earlier you can get their minds wrapped around the financial magnitude of this undertaking, the better. Give them a little time to recover from the staggering figures contained in your budget.

You'll be glad you did this because your very next task is to file a capital or fixed-asset request for the actual funding for the machine. Even if only a short time elapses between submitting the budget and submitting the capital request, you'll be glad you gave the finance people at least a little warning. It goes a long way to making the approval process go more smoothly. Because there are various levels of approval for the request, the process can take between 1 and 4 months to complete.

1.12 Develop a Budget for the Machine Tool Foundation

This is where we begin to discuss the platform or *foundation* upon which the machine tool will be installed. This piece is crucial to the proper operation of the machine. Unless the foundation is designed and constructed to the exact specifications and intended use of the machine tool *and* all the other particulars that are unique to your facility, the machine tool will simply not function as it is intended to. This is why the bulk of the information in this book will address the design and construction of the machine tool foundation. As utterly complex as the machine tool selection and purchase are, designing and building a foundation for that tool that is exactly right for it under all conditions of operation is by far the most difficult part of this project. One small error in design and/or construction of the foundation can render your multimillion-dollar machine tool useless!

The first step toward developing a cost for the machine tool foundation is to secure a copy of the preliminary foundation drawings from the manufacturer. These drawings are crucial to getting both the budgeting and the design processes off to the right start.

The first thing to look for in the drawings is the *foundation depth and width* In addition to the specifics of the machine tool itself, the foundation depth and width depend largely on the conditions of the soil upon which your facility is built. Some facilities are built on hard clay and bedrock. Others, closer to the coasts, are built on sandy soil that is soft and often moist, and there are any number of soil conditions in between these two. The importance of knowing the soil conditions and the allowances you will have to make to accommodate your particular situation cannot be overstated. More on the design and construction of the machine tool foundation will follow in later sections. For now, we are concerning ourselves with creating the budget for the foundation.

As with creating the budget for the machine tool, the budget for the foundation should be prepared as early in the project as possible, and for the same reason: sticker shock! The foundation normally adds to the project a cost approaching 20 to 30 percent of the purchased price of the machine tool itself.

Costing out the foundation is an extremely involved process. Scores of details must be taken into consideration, and to miss any one of them can render the foundation (and therefore the machine tool) inoperable.

1.13   Develop an Electrical Cost of the Machine Tool

Often overlooked in costing out the construction of the foundation is the electrical service. Obviously, machine tools require power to operate—lots of it. Supplying the appropriate power to the new machine tool is a huge part of the foundation cost. A number of factors will play into the cost of supplying power to the machine tool, such as what size service is needed (how much power the machine tool requires) and where in the facility that power will come from (where the nearest electrical service point is).

The power requirements of the machine tool are provided with the manufacturer's specifications. Most machine tools have very high-voltage requirements, such as 480 VAC, 3-phase and sometimes more. What it's actually going to take to get the correct power setup for your machine tool in your facility is information that is best provided by your electrical engineer or facilities electrician.

Unless you have the specific expertise to calculate all this, don't do it alone. Bring in the experts! They can guide you in this complicated part of the budgeting process and make you aware of every necessary detail. For example, if your foundation requires a pit below it, you will need to include a separate 110-VAC (standard household voltage) source around the foundation for lighting the pit. Machine tools often come with access to 110 VAC in their control cabinets, but it is not of sufficient amperage to run lights and power outlets at and under the foundation. Another reason not to use the machine tool control box for lighting is that if you need to turn off or disconnect the machine tool from its power source, you will have no lights or power outlets either. That's why it's always best to put these ancillary power needs on a separate circuit that comes from within your facility.

1.14 Develop a Facility Modification Budget for the Machine Tool

The list of items that comprise this part of the project is long and expensive. Your machine tool will require dedicated supply lines for water (room temperature as well as hot and chilled), compressed air, and coolant, as well as safety rails and a number of other necessary accommodations.

Most machine tools require coolant for their cutting tools. This is supplied in a water-coolant mixture comprised of 93% water and 7% coolant. Based on this, if you are using a 500-gal tank, you will require 460 gal of water. Coolant also has a rapid evaporation rate, and you will quickly lose coolant. This should be closely monitored. Compensating for this coolant loss requires ready access to fresh coolant at the tool site.

The machine tool also uses compressed air—lots of it. Make sure you have a supply line that is equal to the task. If the manufacturer specifies a 1-in compressed air supply line, you will need to install a 2-in line instead. This is so because you will need enough available compressed air to serve the machine tool and other things, such as retractable air reels, special deburring tools, tool clamping machines, and other pneumatic equipment. Additional smaller air supply lines mean you will require a larger main supply line than what the manufacturer specified, which is intended to service just the machine tool.

Chilled water is necessary for the machine tool's heat exchanger. The heat exchanger maintains the optimum oil temperature for the machine tool's use. The most common heat exchangers are the tube style or plate type. Both require chilled water to cool the oil that is heated by friction and by the other heat-generating parts of the machine tool.

Safety rails are an important part of the purchase of a machine tool, and they are often overlooked. Some machine tool manufacturers will incorporate safety rails into the overall design of the machine tool, but these rails are usually much more costly. The buyer is better off to purchase and install them separate from the machine tool purchase. Which safety rails are used often depends on what the company has set as the standard type of safety rail for the entire operation. Whatever the case, it is essential to consult the facility's safety manager on this. Provide her or him with a drawing that illustrates exactly where the safety rails will go.

1.15   Develop a Removal and Demolition Cost for the Old Machine Tool and Its Foundation

In all probability, the machine tool you are buying is to replace an existing one. There is considerable cost in removing the old machine tool. As the project moves forward, the strict time constraints for removing the old machine will quickly become apparent. Sometimes, old machines are still useful and can be sold to used-equipment dealers. In other cases, the machine tool has little or no resale value and must be scrapped. The funds gleaned from scrapping the old tool can often be used to cover the cost of removing it.

If the machine can be resold, contact a used-equipment company to come and remove the machine. You are not in the business of selling equipment, and you do not have the time to find a buyer and get the machine removed in accordance with a tight project schedule. Just get it gone! Let the used-equipment company come get it and be done with it.

Next, the remaining foundation will likely need to be demolished and removed. This is a difficult, costly, and very time-consuming task. In fact, in most cases, it is more cost-effective to choose another location for the new machine tool. Then you can simply remove the old machine tool and cap off the old foundation with concrete and a floor plate, freeing up that space for other uses.

All these things are important considerations in the overall budget cost of the new machine tool.

1.16   Develop a Tooling Cost for the New Machine Tool

Every manufacturing process is unique. It is easy to overlook all the special tooling and attachments needed for your particular process. For example, I worked for a company that purchased a horizontal boring machine. This machine tool was purchased to machine a certain type of very expensive casting. No one in the organization accounted for the funding needed for the special head attachments or tooling needed to finish the machining process associated with this particular part. That head attachment added about 25 percent to the original cost of the machine tool. The rest of the tooling needed added another 15 percent to the total cost. Without these additional components, their multimillion-dollar machine tool would be useless to them.

Of course, these pieces could still be purchased after the fact. But the special pricing we would have received from the manufacturer at the time of the machine tool purchase was no longer available to us, and these needed components would now cost the company significantly more. I'm sure that explaining this very costly omission to upper management was an unenviable task, to say the least. This mistake caused the entire project to be placed on hold for an entire month, until we were able to reassure management that we really did know what we were doing and to regain their confidence.

When considering tooling, you must include all the special fixtures needed, such as large-angle plates, box tables, clamps, box jacks, hairpin clamps, all thread studs, etc.

Much of this equipment can be purchased secondhand from used machine tool suppliers. It helps if you have a purchasing department that will allow you to procure the things you need without a lot of red tape. (Table 1.5 is a list of used machine tool and equipment dealers in the United States.)

Used Machine Tool Dealers	Supplier Country of Origin	City/State in Country
Liberty Machine Tool	United States	Chicago, IL
Prestige Machine Tool	United States	Philadelphia, PA
Yoder Machine Tool	United States	Toledo, OH
Mohawk Machinery Sales	United States	Cincinnati, OH
Machinery Values	United States	New York, NY
Industrial Machinery Sales	United States	Baltimore, MD
Midwest Machine	United States	Chicago, IL
R. P. Machine	United States	Statesville, NC

TABLE 1.5   Used Machine Tool Dealers List

1.17   Final Budget Costs

Table 1.6 is an example of the known budgeted costs associated with the purchase of a large horizontal boring machine. By discussing these costs with your upper management early in the purchase process, you won't catch anyone off guard in the process or later when you create a purchase request for each item. (See Table 1.6 for a sample final budget.)

It will be up to your accounting and operational department to determine if these costs should be labeled as capital or expensed. Depending on your company this will either help the project move to the next phase to purchase the machine tool or cancel the project all together.

1.18   Create the Overall Project Schedule

At this point, it is time to create the first overall project schedule for the machine tool. This document is for the benefit of upper management. This will be the bird's-eye view of the overall project schedule. See an example of the first overall project schedule. (See Fig. 1.1 for a sample schedule.)

Once you and your team have invested all your best efforts into preparing the budget document, the time comes when the budget has to do what it was created to do—persuade the powers that be to give you the money to move forward.

You don't just stuff your budget in a manila envelope and drop it on somebody's desk, hoping it will impress her or him enough to get approval. Your budget has to be presented to upper management *personally*. This can be stressful and intimidating. You and your team have worked harder on preparing this document than perhaps you have on many other tasks, and you all know that if the project does get approved, the hardest work is yet to come.

It's up to the project manager to schedule a meeting with upper management to present the whole budget, item by item. You will be expected to explain each topic and why you budgeted the amount(s) you did. Expect every item to be scrutinized to the nth degree. The people you are presenting to are management and finance people, not from engineering or manufacturing. The best way to persuade them is to educate them. While the meeting will focus on the budget document,

Item	Budget Cost—Estimate	Notes
Basic machine cost (including all taxes, duties, shipping, transportation, etc.)	$1,200,000	Based on First Draft Specification
Options		
Rotary table	75,000	
Chip conveyor	15,000	
Floor plate	35,000	
Machine tool installation	150,000	
Cost of spare parts	10,000	
Tooling	50,000	
Head attachments	50,000	
Warranty (extended)	20,000	
Total budget cost for basic machine and options	1,605,000	
Foundation		
Soil borings and samples	$5,000	
Safety barriers and dust walls	10,000	
Demolition of the existing foundation	50,000	
Cost for design of the foundation	15,000	
Foundation costs	300,000	
Safety rails		
Total budget foundation costs	380,000	
Facility Costs		
Electrical (400amp service)	$40,000	
Low Voltage power (110VAC)	5,000	
DNC and Ethernet (for the IT Dept.)	5,000	
Isolation Transformer	7,500	
Compressed Air	7,500	
Chilled Water	7,500	
Process Water	5,000	
Total Budget Facility Costs	77,500	
Total Overall Budgeted Costs	$2,062,500	
20% contingency	2,475,000	

TABLE 1.6 Budget Cost for the Machine Tool HBM Based on First Draft of Technical Specification

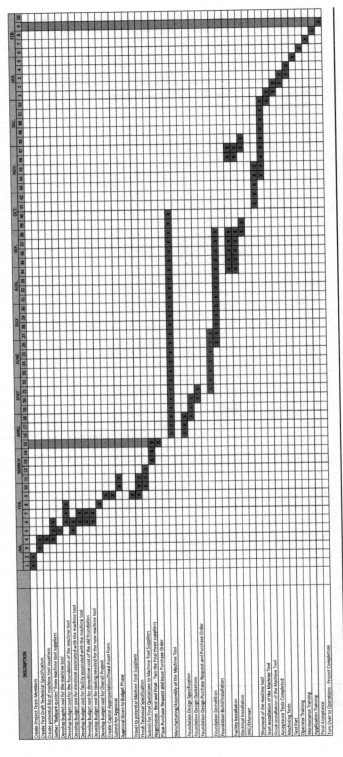

FIGURE 1.1 Budget phase project schedule.

the most important thing to that room full of decision makers is your credibility and the credibility of your team. They understand the numbers. They need to know that because of your collective diligence and expertise the numbers are an accurate depiction of what it's going to take to purchase this improvement and make it do what you say it's going to do for the company's bottom line.

The presentation process is an invaluable education in itself. It allows you to understand more clearly what is and is not important to your peers and to the higher-ups. It always amazes me what kind of questions people think are important. This is the most important question you will be asked: when would the machine tool project be completed? That's really the main thing that upper management wants to know, and if they're asking that question, it's a very good sign.

1.19   Capital Appropriation and Level of Authority

Once your budget is approved, your project is ready to request funding. You will need to file the proper forms your company uses to fund capital or fixed-asset requests. Naturally, there is a process for funding capital purchases in every company, and it always involves the highest levels of management. Patience is a virtue, as this process can often take a long time.

1.20   Summary

By now, you and your team are probably very tired of all this planning. But as the old saying goes, *Failing to plan is the same as planning to fail.* In the case of the selection, purchase, and installation of a multimillion-dollar machine tool, this is especially true. Whether the outcome is what was originally desired or tragically falls short of expectations (and tragic it would be) depends largely on how well and how thoroughly your team has done their homework and communicated it to all concerned in a clear and complete fashion. In spite of all this careful planning, sometimes the management team will decide to pull the plug on the whole project anyway. It just happens from time to time. Unforeseen variations in cash flows, strategic reconsiderations, and company policy changes are just a few of the things that can affect whether a project like this will ever get past the budgeting stage. Hopefully, your project will move along as planned to its swift and successful completion.

In this chapter, we covered the basics of budgeting for your machine tool project. It is not always this clean and tidy. Each situation is different, and you will have to learn how to compensate for the various distinctive qualities of your particular project when applying the methods outlined in this chapter. I hope that this information will make the whole process easier.

CHAPTER 2
Procurement

2.1   Prespecification Meeting

As we emphasized previously, the most important stage in preparing to purchase a machine tool is the technical specification. This specification will articulate the requirements and options the team has identified as essential to choosing the right machine tool for the company. All other planning and documentation will ensue from this document. Table 2.1 is a list of possible team members based on their departments and responsibilities within a manufacturing company.

Preparing the technical specification will be more of a process than an event. It isn't something that comes together neatly, stage by stage, until it's completed. Rather, think of it as a living document that evolves as your planning process evolves. It will require numerous changes and revisions before the process is complete.

Naturally, the first step is to get the entire team together and review the draft technical specification you prepared in the budget phase just completed. I recommend that you schedule this meeting on for a Monday before everyone's minds are cluttered with the things being worked on that week. The meeting, if conducted correctly, will likely require the entire day, so let everybody know not to schedule anything else. The project manager should make sure lunch and breaks are provided to ensure that the team remains focused on the task ahead. I understand that it is almost impossible to get 10 or 15 people together every week for a meeting, but that is required when purchasing a large machine tool.

Department	Team Member #1 Choice	Team Member #2 Choice
Operations/ Manufacturing	Manufacturing engineer Machinists	Crane operator Operations supervisor Machinists
Purchasing	Capital purchasing agent	Purchasing agent (MRO) Purchasing manager Purchasing agent
Facilities	Electrical facilities engineer Mechanical facilities engineer Plant engineer	Electrician Plant technician Facilities manager
Maintenance	Controls engineer Mechanic Electrician	Maintenance supervisor Maintenance manager Maintenance engineer
Accounting	Accountant	Controller
Engineering	Machining engineer	Engineering manager
Programming	NC programmer	Engineering manager
Quality	Quality engineer	Quality manager

Table 2.1   Project Team Members

I omitted a few departments within the manufacturing environment in Table 2.1, but they are almost equally important to the overall success of the project. They are as follows:

- Environmental, Health, and Safety (EHS)
- Shipping
- Receiving
- Transportation
- Warehousing

These departments are again equally important and should be supplied with any and all information related to the purchase of the machine tool. For example, the receiving department needs to know that your team has purchased a large machine tool and that when the machine tool arrives at your facility, it should not be off-loaded in the receiving dock. If it is, that will cause you to make additional crane or forklift moves within your facility and will waste your employees' time and create a possibility of damage to the equipment by making extra moves.

2.2   Machine Tool Project Specification Document

This is the beginning of creating the first draft of the technical specification. At this point, neither you nor your team will know exactly what machine tool your company needs. This is where you all begin working together to figure that out. Managers, engineers, supervisors, and machinists will all be asked to weigh in. Every question is important and every opinion matters.

The process may go something like this: The manufacturing and tooling engineer defines the requirements for the type of material holder the company needs for this type of machine. The facilities engineer explains the type of incoming power needed to run the machine. The computer numerical control (CNC) programmer and the operations people recommend the specific type of CNC that best fits the needs of the company, manufacturing department, and project. In these three examples alone, many choices will be available. It will require plenty of discussion to arrive at what seem to be the right choices at this early point in the process.

What you envision on the first day will likely be very different from what you end up with when the leadership decides that it has the specification it's going to move forward with. Even then, new information about options, costs, and anything else will usually present itself, requiring some last-minute tweaking.

2.3   Selection of the Project Leader for the Entire Project except the Foundation

Every project needs a leader. This is the time to decide which team member is right to lead the team to the best possible outcome in the purchase of the right machine tool. This choice does not involve the foundation design specification or the machine tool installation specification. That comes later and will require the same kind of planning and similar technical specification as the team is addressing now.

An increasing number of companies have a dedicated project engineer whose purpose is to lead these large, time-intensive projects. However, this practice is still more the exception than the rule. If your company does not have such a position, the manufacturing engineer within the department that will use the machine tool and the maintenance and facilities engineer are the most logical considerations. In my experience, of these two, the preferred choice is the maintenance and facilities engineer. This person has the expertise to both lead this complex undertaking and plan and install the foundation. The maintenance and facilities engineer can coordinate all the electrical, air, water, and other needed connections for the proper installation and operation of the new machine tool. In other words, this person can take the whole project from the earliest planning stages up to the point where the foundation design specification and machine tool installation need to be addressed.

Still, if you have the benefit of a project engineer, that person's specialty is project management of this type. He or she can lead all phases of planning and, when it's time to start working the plans, can see it through from start to finish. The project engineer works at ground level with the project, making sure the day-to-day functions move forward in a timely fashion and that everything is done according to the plan.

Another advantage to having a dedicated project engineer is that the two others mentioned, the manufacturing engineer and the maintenance and facilities engineer, cannot devote all their time exclusively to one project—and there is usually more than one project in process at a time. The maintenance and facilities engineer is kept quite busy with other kinds of projects more specific to her or his area and in handling the never-ending spate of problems that arise during normal daily operations. The project engineer has only one responsibility at any given time. In this case, it's this project. The project engineer isn't the person to manage the foundation piece of the project, but should be on the team doing the foundation design specification and machine tool installation to monitor its progress.

The larger and more complicated the purchase of a machine tool is, the better it is for the project engineer or manager to be the overall leader of the project. Also, the larger the project, the greater the likelihood that it's the leader's only project.

2.4   Machine Tool Supplier Visits and Quotes

This is the appropriate time for the entire team to begin making appointments with machine tool suppliers (manufacturers) to gather all the information available about the equipment and available options. It's also not too early to talk prices.

Find out which supplier has the best support team. Which one has been in the business the longest? Which company is the most fiscally sound? Which company is the leader in the industry? (See Table 2.2 for a list of questions that you and

Question	Answer
Does the machine tool supplier have a service provider in the United States?	
Where is the physical location of the U.S. machine tool supplier?	
Where is the training location in the United States?	
How many horizontal boring machines (HBMs) are there worldwide?	
How many HBMs are in the United States?	
How many years has the machine tool supplier been operating in the United States?	
What is the number of service personnel in the United States?	
What is the total number of employees in the United States?	
What is the total number of employees worldwide?	
What are total sales worldwide?	
What are total sales in the United States?	
Will the machine tool be manufactured in the United States? If not, in what country?	
Will the machine tool be assembled in the United States? If not, in what country?	
What is the value of spare parts in inventory?	

TABLE **2.2**   Questions for the Machine Tool Supplier

your team should ask before deciding which machine tool supplier is the best choice.)

Once the team has completed its fact-finding mission, the entire team should review every first quote they have received from each supplier. Quotes will differ substantially, both in items listed and in pricing. This is so because machine tools themselves differ greatly from manufacturer to manufacturer. Table 2.3 details both a rating scale for all machine tool manufacturers and your and your team members' opinions of each machine tool supplier.

The team's task is to sift through all the quotes to determine how they vary. This will require more meetings and plenty of time spent gathered around a computer. It then becomes the project manager's responsibility to present a detailed summary of the differences and similarities between the quotes and of the machine tools and equipment each represents. Table 2.3 details the differences among all the machine tools that were quoted for your facility.

2.5   Machine Tool Supplier References

Every machine tool supplier you are considering will expect you to ask for references. This creates additional opportunities to gather information on the suppliers and their products. What's more, this is a great opportunity to request permission to visit the reference's facility and see how the equipment, as well as the entire operation, works. You may discover that the machine tool the team is leaning toward isn't quite as good as originally thought. Maybe it is. Other people in businesses that buy machine tools are usually very candid in their comments about the equipment they've purchased and the supplier from whom they purchased it. They will readily tell you what they like and don't like about both. You can learn firsthand how the machine tool functions and see it in action. You will learn about its strengths and its limitations—its impressive qualities and its quirks.

Other invaluable information you can glean from a visit concerns how the installation went at that facility. What problems did they ran into? What insights can they offer from the installation that could save your company a great deal of time and money?

Horizontal Boring Machines Suppliers

Supplier Country of Origin	Does the Machine Tool Supplier have a Service Provider in the U.S.?	Cost	Cost Rating	Service	Number of U.S. Employees	Name Brand	Number of HBMs in the Marketplace	Number of HBMs in the U.S. Marketplace	Years of Service in the U.S.
Machine Tool Supplier									
	Yes	High		Good	500	Good			
	No	Low		Fair	NA	Fair			
	Yes	Medium		Good	150	Good			
	Yes	High		Poor	25	Good			
	Yes	Medium		Fair	50	Good			
	Yes	Medium		Fair	50	Good			
	Yes	High		Good	300	Best			
	Yes	High		Fair	25	Good			
	Yes	Medium		Fair	30	Good			
	Yes	Low		NA	None	Good			
	Yes	Medium		Fair	100	Good			
	Yes	Medium		NA	NA	Good			
Rebuilder of Used HBM's									
	Yes	Medium		Good	60	Good			
	Yes	Medium		Good	120	Good			
	Yes	Fair		Good	30	Fair			
	Yes	Fair		Best	10	Good			
	Yes	Medium		Fair	40	Fair			

TABLE 2.3 Rating Scale

Sometimes the machine tool supplier will have one of its representatives arrange the visit and accompany the team to the reference's facility. This is common and not objectionable. However, the project leader should make it clear to the sales representative that some private time between the reference and the team is expected. Nothing should prevent the reference from speaking freely about his experiences with the machine tool, the supplier, and even the sales representative.

Once the team has obtained the quotes, done all its fact-finding, and agreed on which machine tool best meets the company's needs, it's time to compare all three machine tool quotations in spreadsheet form. After you review the quote comparison of the three suppliers, this information will lead you to finalize and complete the first draft of the technical specification. Then engineering, manufacturing, and facilities maintenance should all sign off on it.

Table 2.4 shows a comparison of at least three machine tool suppliers' quotations and the differences in specifications among the machine tool suppliers. This example is not far-fetched. There will be extreme differences between each machine tool supplier.

I recommend that you try to contact at least 7 to 10 machine tool suppliers for the given machine tool that your company is looking for. Even though the total machine cost among the three proposed machine tools is less than $500,000, you can see there is a big difference. The smaller the differences among the three quotations, the better overall job you did in selecting the machine tool supplier and detailing the specification. The machine tool supplier that you choose will drive the final approval of the machine tool specification.

2.6   The Final Approved Machine Tool Specification

After selection of the machine that you want to purchase, it is extremely important that you update the final approved machine tool specification with the details of the final specifications and items that you and your team would like to purchase. This final approved machine tool specification basically is the specification that the purchasing group will need to send out to the preferred and selected machine tool supplier from

Horizontal Boring Machine—Floor Type Including Table			
Item	**Company ABC**	**Company RTS**	**Company XYZ**
X-axis travel (longitudinal)	100"	120"	144"
Y-axis travel (vertical)	78"	96"	96"
Z-axis travel (boring/quill)	24"	36"	39.37"
W-axis travel	No ram included	24"	24"
B-axis contouring	Yes	Yes	Yes
V-axis travel	36"	36"	39.37"
X-axis construction	Linear guideway	Boxway construction	Linear guideway
Y-axis construction	Linear guideway	Boxway construction	Linear guideway
Spindle horsepower	30 hp	50 hp	60 hp
Spindle thrust	1200 ft/lb min.	1200 ft/lb min.	1600 ft/lb min.
Spindle diameter	5"	5.1"	6"
Spindle tool taper	CAT 50/50 taper	CAT 50/50 taper	CAT 50/50 taper
Number of spindle gear ranges	2	2	2
Spindle rpm	0–800 rpm and 750–1500 rpm	0–1000 rpm and 1000–1800 rpm	0–800 rpm and 750–2000 rpm
Coolant tank size	250 gal	400 gal	500 gal
Through the spindle coolant	Yes	Yes	Yes
Number of coolant connections at front of ram	6	6	6
Minimum clamping force of the retention knob/tool	6600 lb/force per design	5500 lb/force per design	7000 lb/force per design
M19/spindle orientation included	Yes	Yes	Yes
Work light included	Yes	Yes	Yes

TABLE 2.4 Quote Comparison

Horizontal Boring Machine—Floor Type Including Table			
Item	Company ABC	Company RTS	Company XYZ
Size of ram	None	12" × 12" square ram	12" × 16" rectangle square ram
Size of rotary table	48" × 48"	48" × 60"	60" × 60"
Weight capacity of table	20,000 lb	30,000 lb	40,000 lb
Locking pin for rotary table every 90°	Yes	Yes	Yes
B-axis construction	Linear guideway	Boxway construction	Linear guideway
V-axis construction	Linear guideway	Boxway construction	Linear guideway
Rapid traverse speeds			
X-axis	400 ipm	600 ipm	400 ipm
Y-axis	400 ipm	600 ipm	400 ipm
Z-axis	300 ipm	500 ipm	300 ipm
B-axis	360° per minute	360° per minute	720° per minute
V-axis	300 ipm	500 ipm	300 ipm
Maximum cutting speed (all axes)	150 ipm	175 ipm	150 ipm
Telescopic way covers included	Yes	Yes	Yes
Control options	Fanuc 31I	Siemens 828D	Siemens 828D
Tool changer	60-position tool changer	90-position tool changer	90-position tool changer
Max. weight of tool	55 lb	75 lb	100 lb
Feedback system	Heidenhain incremental glass scales	Heidenhain Incremental tape scales	Fagor Incremental tape scales
Repeatablility of all axes	0.0001"	0.0001"	0.0001"
Accuracy of rotary table	±3.5 arc seconds (min.)	±3.5 arc seconds (min.)	±3.5 arc seconds (min.)
Color of the machine tool	Dark gray	Dark blue	Gray

TABLE 2.4 Quote Comparison (*Continued*)

Horizontal Boring Machine—Floor Type Including Table			
Item	Company ABC	Company RTS	Company XYZ
Installation included?	Not included	Not included	Included
Supply voltage, three-phase	460 VAC/three-phase/(±10%)	460 VAC/three-phase/(±10%)	460 VAC/three-phase/(±10%)
Isolation transformer included?	No	Yes	Yes/No
Training			
Operator	One week of training included (4.5 days total)	One week of training included (4.5 days total)	One week of training included (4.5 days total)
Maintenance	One week of training included (4.5 days total)	One week of training included (4.5 days total)	One week of training included (4.5 days total)
Application	One week of training included (4.5 days total)	One week of training included (4.5 days total)	One week of training included (4.5 days total)
Spare parts list	Recommended spare parts list included	Recommended spare parts list included	Recommended spare parts list included
Spare parts included?	No	Yes	Yes
Warranty	12 months	24 Months	12 Months
Terms and conditions	20% down with purchase order	30% down with purchase order	25% down with purchase order
	30% with sign-off at the machine tool factory	30% with sign-off at the machine tool factory	25% with sign-off at the machine tool factory
	30% with delivery of machine tool at buyer's factory	30% with delivery of machine tool at buyer's factory	25% with delivery of machine tool at buyer's factory
	20% four weeks after acceptance of machine tool at buyer's factory	10% four weeks after acceptance of machine tool at buyer's factory	25% four weeks after acceptance of the machine tool at buyer's factory

TABLE 2.4 Quote Comparison (*Continued*)

Horizontal Boring Machine—Floor Type Including Table			
Item	Company ABC	Company RTS	Company XYZ
Leveling bolts/ wedges/anchor bolts included?	Yes	Yes	Yes
Acceptance/ alignment/laser shots included?	Yes	Yes	Yes
Machine runoff/ horsepower test (min. 75%) cut	Yes	Yes	Yes
Detailed list of G and M codes for this machine tool	Yes	Yes	Yes
Manual operation mode (manual feed/spindle on)	Yes	Yes	Yes
Detailed assembly drawings of all major machine components	Yes	Yes	Yes
Detailed drawings of all reusable parts	Yes	Yes	Yes
Cost of Basic Machine	**$775,995**	**$850,000**	**$1,220,000**
Options			
Chip conveyor	$25,000	$30,000	$45,000
Chip deflectors	$10,000	$7,500	$17,000
Right-angle head	$125,000	$115,000	$200,000
Universal head/ manual adjustment	$100,000	$95,000	$150,000
Universal head/ automatic adjustment	$200,000	$175,000	$350,000
Cost of machine installation	$175,000	$150,000	$0
Cost of spare parts	$65,000	$0	$0
Total cost (including options)	$1,475,995	$1,422,500	$1,982,000

TABLE 2.4 Quote Comparison (*Continued*)

whom you want to purchase the proposed machine tool. Therefore the purchase order by the purchasing group will reference the final approved machine tool specification in the details of the purchase order. It is also important that you get sign-offs and signatures of the appropriate department heads of operations, maintenance and facilities, engineering, projects, and purchasing before the final purchase order is sent to the machine tool supplier.

2.7 Checklist for the Project Leader/Manager during the Procurement Phase

During the procurement phase, it is the responsibility of the project leader or manager to ask the simple questions related to the purchase of the new machine tool. These questions are simple but should be asked at the beginning of every machine tool purchase, and they are detailed in Table 2.5.

2.8 Operator Training

Included in the cost of the machine tool is at least one week of operator and maintenance training not only at the machine tool supplier's facility but also in the factory or plant site after the installation is completed. This is a must. As the project leader, you should make this one of your top priorities. Make sure you add this training into the purchase price of the machine tool. Training is another way to allow all team members of the project to "buy in" to the idea of purchasing a new machine tool. I would also recommend at least 1 week of application engineering and programming training for the NC programmers and machining engineers. See the Table 2.6.

2.9 Maintenance and Facilities Training

Maintenance training is also imperative when you are purchasing a machine tool. A different type of person is needed for servicing a machine tool than is for running it. Table 2.7 is a checklist of items that your maintenance department electricians and mechanics should receive if you purchase a machine tool. See Table 2.7 for the detailed maintenance training checklist.

Department	Responsible	Question	Sign-off and Date
Purchasing	Purchasing agent or manager	What are the standard terms and conditions for the purchase?	
Purchasing	Purchasing agent or manager	What are the payment terms for the purchase?	
Facilities and maintenance	Electrical engineer or electrician	What is the standard three-phase and single-phase VAC power at your facility?	
Facilities and maintenance	Electrical engineer or electrician	Where is the closest power source in relation to the proposed machine tool location?	
Facilities and maintenance	Electrical engineer or electrician	Where is the closest pneumatic/compressed air connection for the machine tool?	
Environmental, health, and safety (EHS)	Safety manager	Has the safety manager reviewed the proposed machine tool location?	
Environmental, health, and safety (EHS)	Safety manager	Has the safety manager reviewed the proposed machine tool safety functions?	
Environmental, health, and safety (EHS)	Safety manager	Has the safety manager reviewed the proposed machine tool safety rails, and does the machine tool have any confirm spaces?	
Operations/ manufacturing	Manufacturing engineer	Was the machine tool or piece of equipment laid out on the factory floor? Is there a plan and elevation drawing?	
Operations/ manufacturing	Manufacturing engineer	Was the location of the machine tool or piece of equipment verified by the maintenance and facilities department for proper placement?	
Operations/ manufacturing	Machinist and operational supervisor	Were the machinist and operational supervisor notified that a new machine tool was being purchased?	

TABLE 2.5   Checklist for the Project Phase at the end of the Procurement Phase

1. *Turning on the power on the machine tool.* It's important for the operator to understand how to start up the machine or machine tool from a power off condition. It's also a good idea for the operator to understand how to properly turn off the machine in a nonemergency condition. Almost every machine tool will have a different set of start-up and shutdown procedures. You as the project leader will need to create a procedure for both.

2. *Starting up the CNC.* After the proper power on condition is achieved, start up the CNC.

3. *Resetting the machine after an emergency stop.* The next step is to reset the machine tool from the emergency stop condition. During this step ask the machine tool supplier to detail the safety precautions of the machine tool.

4. *Resetting the CNC error messages.* Ask the machine tool installer to show the operator how to reset the CNC after an alarm condition. Also request that the machine tool supplier list all special machine alarms and how to recover from each fault.

5. *Referencing the CNC.* This is the process of having the CNC find the home position. You will have to home all the axes before you can run the machine tool in automatic. In this process the operator must be shown how to reference each and every axis. Most of the newer machine tools today have distance-coded scales. Therefore the operator will have to move only approximately 40 to 80 mm to reference. Some machine tools automatically are referenced when there is no fault on the CNC during power-on.

6. *Putting the CNC in manual.* It's a simple task, but the operator has to understand how to move the machine tool in manual mode. Sometimes this is called the *jog mode*. Also ask the machine tool supplier to show your operators how to move the machine in rapid.

7. *Tool change.* The operators must be able to perform a tool change, if the machine tool has that option, by themselves. The standard ISO/EIA code for a tool change is M06.

8. Loading a CNC program into the control and putting the machine tool in automatic and manual data input (MDI) mode.

9. *Program directory.* Most of the newer machine tools and CNCs are Windows PC based and therefore have some type of program storage directory. The machine tool supplier should show the operator and engineer how to create and save a program and to make a new directory in the CNC.

10. *Punching out a CNC program.* Punching out a program occurs when you actually output a CNC program to your PC or some other storage device.

TABLE **2.6**   Operator Training Checklist

11. Detailing all the oil reservoirs and what kind of oil is needed in each tank.
12. Detailing all the oil filters for each tank.
13. *Work offsets and tool offsets.* Ask the machine tool installer to show you how to perform both operations.
14. *Using the handheld pendant.* Every large CNC machine tool will have a handheld pendant (HHP) or handheld unit (HHU). Make sure that the machine tool supplier shows your machinist and engineer every button and key on this unit. In almost every new installation that I have been involved with, the machine tool supplier had to fix something on the HHU.
15. *Copying a CNC program.* Many times in the machining process you need to copy a program or sections of a program.
16. *Backing up the CNC memory.* Normally this is performed by the maintenance and facilities group; but if you don't have this department within your organization, it will be up to you to back up this information.
17. *Reposition.* During automatic or MDI if you interrupt the program, ask the machine tool supplier to show you how to reposition back to the interruption point.
18. *Searching within the CNC program.* Search by block or G code.
19. *Special G codes for this machine tool.* Every machine tool has its own special G codes. Ask the trainer to go over each special G code.
20. *Special M Codes for this machine tool.* Every machine tool has its own special M codes. Ensure that the trainer goes over each special M code.
21. *Modal G and M codes.* The G and M codes are active when the CNC starts up.
22. *Reset condition.* When the operator presses the reset button on the control, how does it affect the CNC part program?
23. *Loading/unloading of the tool in the spindle.* Ask the machine tool supplier to show each and every operator and maintenance personnel what type of retention knob their machine tool needs.
24. Explanation of the tool probe and how the data are input into the CNC.
25. Explanation of the relative and absolute coordinate systems in Fanuc CNCs and the work and machine coordinate systems in the Siemens CNCs.
26. *Note:* Refer to the machine tool manufacturer for additional operator-related training items.

TABLE 2.6 Operator Training Checklist (*Continued*)

1. *Turning on the power to the machine tool.* It's important for the operator to understand how to start up the machine or machine tool from a power off condition. It's also a good idea for the operator to understand how to properly turn off the machine in a nonemergency stop condition. Almost every machine tool will have a different set of start-up and shutdown procedures. You as the project leader will need to create a procedure for both.

2. *Starting up the CNC.* After the proper power on condition is achieved. start up the CNC.

3. *Resetting the machine from an emergency stop.* The next step is to reset the machine tool from an emergency stop condition. During this step ask the machine tool supplier to detail the safety precautions of the machine tool.

4. *Resetting the CNC error messages.* Request that the machine tool installer show the operator how to reset the CNC after an alarm condition. Also ask the machine tool supplier for a list of all special machine alarms and how to recover from each fault.

5. *Referencing the CNC.* In this process the CNC finds the home position. You will have to home all the axes before you can run the machine tool in automatic. In this process the operator must be shown how to reference each and every axis. Most of the newer machine tools today have distance-coded scales. Therefore operators have will have to move only approximately 40 to 80 mm to reference. Some machine tools automatically are referenced when there are no faults on the CNC during power-on.

6. *Putting the CNC in manual.* It's a simple task, but the operator has to understand how to move the machine tool in manual mode. Sometimes this is called *jog mode.* Also ask the machine tool supplier to show your operators to move the machine in Rapid.

7. *Using the handheld pendant.* Every large CNC machine tool will have a handheld pendant (HHP) or handheld unit (HHU). Ensure that the machine tool supplier shows your machinist and engineer every button and key on this unit. In almost every new installation that I have been involved with, the machine tool supplier had to fix something on the HHU.

8. *Backing up the CNC memory.* Store the backup on the CNC and on your personal storage device.

9. Backing up the data on the variable-frequency drives (VFDs) and the spindle drive data.

10. *Punching out a CNC program.* Punching out a program occurs when you actually output a CNC program to your PC or some other storage device.

11. Detailing all the oil reservoirs.

TABLE 2.7   Maintenance Training Checklist

12. *Detailing all the oil filters for each tank.* Mark each filter and cartridge with the type of filter that should be used for each.
13. *Marking each oil reservoir with the type of oil required.* It is my opinion that the machine tool supplier should perform this step. But if this has not been done, you must do it during the maintenance training. Make a label or sign, and detail what type of oil should be used for each and every oil reservoir.
14. Detailing all the lubrication points for all bearings that need external lubrication.
15. *Maintaining telescopic way covers.* It is always a good idea to clean the metal chips and lubricate the way covers with WD-40 at least once per month.
16. *Using a password to make changes to the machine control parameters.* If the machine tool does not have a password, you must create one.
17. Making changes to the machine control parameters.
18. *Understanding the electrical drawings.* Your maintenance department must be able to read the electrical drawings and to search within the drawings. Know the general layout of the electrical cabinet.
19. *Understanding the mechanical drawings.* Your maintenance department must be able to read the mechanical drawings and search within the drawings.
20. *Locating the main power panel and making sure it is labeled properly.* Make sure that you are able to lock out the main power for maintenance purposes.
21. *Axis compensation.* Make sure that the machine tool supplier shows you how to make changes to the axis compensation parameters which will be needed for laser calibration.
22. *Software limits of each axis.* Engage each software limit, and make sure that the machine will move off the limit in each direction. Most of the newer CNCs have multiple software limits that can be set up.
23. *Hardware limits of each axis.* Engage each hardware limit on each axis, and make sure that the machine tool goes into emergency stop. Most of the newer CNCs and machine tools don't have hardware limits anymore.
24. Go over all OEM generated alarms and how to recover from each alarm. You will have to specify this in the purchase of the machine tool because this will be very time-consuming and will take at least one or two days.
25. *Monitoring the programmable logic controller (PLC) of the machine tool.* This includes how to search within the PLC for inputs, outputs, timers, counters, and memory bits.

TABLE 2.7   Maintenance Training Checklist (*Continued*)

26. *Viewing the diagnosis screens for each axis.* The diagnosis screen should show the following error and all other axis parameters.
27. *Manual operation of the automatic tool changer.* Detail step by step how to move the automatic tool changer.
28. Recovering after the automatic tool changer hangs up.
29. Ask the machine tool supplier to make a step-by-step procedure of the automatic tool changer operation and detail how to get to the control screens for the automatic tool changer.
30. *Loading/unloading of the tool in the spindle.* Request that the machine tool supplier show each and every operator and maintenance personnel what type of retention knob their machine tool needs.
31. *Known force in pounds per square inch of the spindle drawbar.* Ask the machine tool supplier to not only check and record this value but also compare it with the drawing of the spindle.
32. *Note:* Refer to the machine tool manufacturer for additional maintenance-related training items.

TABLE 2.7   Maintenance Training Checklist (*Continued*)

2.10   Application and Programming Training

Application and programming training is also imperative when you are purchasing a machine tool. The application and programming training can be attended by the manufacturing engineer, machining engineer, programmer engineer, or even the machinists who will run the machine tool. Mainly this training will be needed by the engineering and programming groups. Table 2.8 is a checklist of items that your engineering and programming departments should receive if you purchase a machine tool programming.

2.11   Schedule for the Machine Tool Build

Once the specification is completed, it's time to ask the supplier to prepare a build schedule. The schedule should include all specifics concerning the start date of the build, start the machine tool assembly, final alignments of the machine tool at the machine tool suppliers facility, final acceptance of the machine tool at the machine tool suppliers facility, and finally the shipment of the machine tool to the buyers facility. (See Fig. 2.1)

1. *Detailed list of M and G codes.* Make sure that the machine tool supplier details the entire special list of M and G codes associated with the machine tool.

2. *Tool probe operation.* Ask the machine tool supplier to demonstrate the tool probe operation and all the associated cycles that relate to the tool probe.

3. *Backing and changing the tool probe cycles.* Request that the machine tool supplier demonstrate how to upload and download all the cycles needed for tool probe operation.

4. Have the machine tool supplier detail the header information for the part program needed so that the control will accept it during downloading.

5. Have the machine tool supplier detail the RS232C and Ethernet parameters for uploading and downloading of the part program.

6. *Have the machine tool supplier demonstrate all ways of uploading and downloading of the part programs.* This is important. If the Ethernet system is down for maintenance or just not working for some reason, you have a way to get the part program into the machine control. Normally the newer CNCs will have a USB or PCMCIA port located at the operator's machine control panel.

7. *Have the machine tool supplier demonstrate and document the uploading and downloading of the part program for the programmer and operator.* Have this procedure documented by the programmer, operator, and maintenance.

8. Included in the setup of the machine is the machine tool supplier's help with setting up of the RS232C or Ethernet parameters for uploading and downloading of the part program and cycles associated with the machine tool.

9. *Ask the machine tool supplier to review the supplied programming manual.* This is only for the special functions that the machine tool supplier added to the control. The standard control manuals should be supplied by the machine tool supplier.

10. *Tool and work offsets operation and function.* Have the machine tool supplier demonstrate how the tool and work offsets function on the CNC.

11. *Standard cycles.* Request that the machine tool supplier demonstrate how the standard cycles work on the machine tool.

12. *Using the handheld pendant.* Every large CNC machine tool will have an HHP or handheld unit (HHU). Make sure that the machine tool supplier shows your machinist and engineer every button and key on this unit. In almost every new installation that I have been involved with, the machine tool supplier had to fix something on the HHU.

TABLE 2.8 Application and Programming Training Checklist

13.	*Copying of a CNC program.* Many times in the machining process you need to copy a program or sections of a program.
14.	*Backing up the CNC memory.* Normally this is performed by the maintenance and facilities group; but if you don't have this department within your organization, it will be up to you to back up this information.
15.	*Reposition.* During automatic or MDI if you interrupt the program, ask the machine tool supplier to show you how to reposition back to the interruption point.
16.	*Searching within the CNC program.* Search by block or G code.
17.	*Modal G and M codes.* The G and M codes are active when the CNC starts up.
18.	*Reset condition.* When the operator presses the reset button on the control, how does it affect the CNC part program?
19.	*Loading/unloading of the tool in the spindle.* Ask the machine tool supplier to show each and every operator and maintenance personnel what type of retention knob their machine tool needs.
20.	Explain the tool probe and how the data are input into the CNC.
21.	Explanation of the relative and absolute coordinate systems in Fanuc CNCs and the work and machine coordinate systems in the Siemens CNCs.
22.	*Note:* Refer to the machine tool manufacturer for additional operator-related training items.

TABLE 2.8 Application and Programming Training Checklist (*Continued*)

2.12 Terms and Conditions

Terms and conditions will vary from machine tool to machine tool and from supplier to supplier. Also, the more expensive the machine, the greater the number of stipulations that will apply. The terms and conditions of a purchase are usually not as complicated as one might imagine, but whatever the case, it is the project manager's responsibility to negotiate it all. If there is a problem, it is up to the project manager to solve it. Your company will have its own financial policies regarding purchasing, and those policies will contain the terms and conditions acceptable for major capital purchases. Likewise, the supplier will have its own policy about the terms and conditions of a sale. Often, these requirements differ significantly. It is up to the project manager to reconcile the two and broker a deal that works for both parties.

Since the project manager represents the buyer, and since the buyer usually holds all the cards, the pressure to relax certain

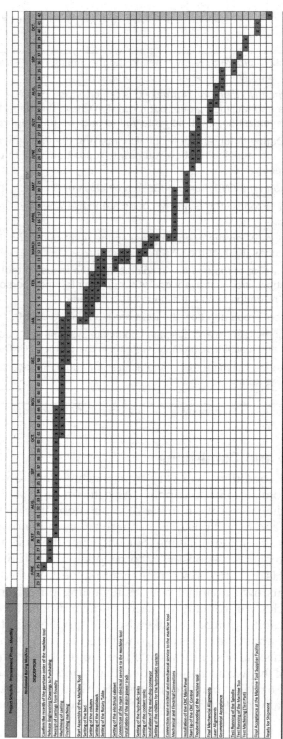

FIGURE 2.1 Procurement phase project schedule.

requirements should be applied to the seller as much as possible. Your finance people won't appreciate being asked to bend too much if they feel the supplier should be making most of the concessions.

Remember that the supplier doesn't make any money if she or he is not selling machine tools. At the same time, respect for the supplier's policies shows you understand he or she has a business to run and that you're dealing in good faith. That said, the supplier needs to make that sale. Make the supplier earn it!

It is essential that the payment terms be as detailed as possible. There should be no gray areas. Every detail must be clearly defined and understood. If there is any uncertainty, get clarification in writing. See Table 2.9 for a sample of terms and conditions.

20 Percent Down with Receipt of the Buyer's Purchase Order

A down payment of 20 percent down is substantial at the beginning of this purchase in which the machine tool supplier gets to invoice 20 percent of the purchase of the machine as soon as the purchase order is placed. In today's manufacturing world, it takes at least 45 to 60 days after the invoice is approved before the machine tool supplier is paid. Let the machine tool supplier know this when the purchase order is placed. Be sure that you let your purchasing and accounting department know that you have approved this 20 percent payment.

30 Percent with Sign-Off at the Machine Tool Supplier's Factory

During this step, we force the two companies to come together to review the machine tool at the machine tool supplier's facility so that you, the customer, is convinced that you receive all the standard equipment and options associated with the

20% down with the purchase order
30% with signoff at the machine tool manufacturer
30% with delivery of machine the tool at the buyer's facility
20% four weeks after acceptance of the machine tool at the buyer's facility

TABLE **2.9**   Standard Terms and Conditions for a Machine Tool Purchase

purchase order. Be sure that you let the purchasing and accounting department know that you have approved this 30 percent payment.

30 Percent with the Acceptance of the Machine Tool at the Buyer's Facility

Before you approve this step in the process, the machine has to be completely installed and the entire geometric acceptance tests completed. Normally there is a sign-off requirement during this step. Be sure that you let your purchasing and accounting department know that you have approved this 30 percent payment.

20 Percent Four Weeks after Acceptance of the Machine Tool at the Buyer's Factory

In this step, you force the machine tool supplier to wait an additional 4 weeks after the third step because it gives you and your company some run time on the machine tool, to see how well it performs. The main reason that I added this section to the book is the fact that in each and every machine tool purchase that I have been associated with, there have been some issues that linger after the installation of the machine tool. This last step forces the machine tool supplier to fix the bugs or issues with the machine as quickly as possible because they supplier wants to get paid.

2.13   Liquidated Damages due to Late Delivery

Liquidated damages due to late delivery of the machine tool are something that varies from machine tool to machine tool. The terms of the liquidated damages can be driven by the cost, installation schedule, or machining performance of the machine tool. The bottom line is the fact that you need to tie something in the purchase of the machine tool to the liquidated damages portion of the contract. The good thing is that it's at the point in the purchase process where you need to visit the machine tool builders and make sure they are on schedule.

I want to stress that you have liquidated damages in the purchase of the machine tool to protect your company. Based on my experience, I've seen liquidated damages in the 25 percent range depending on how critical the project is and how effective your purchasing department is at getting the machine tool supplier to agree to the terms.

2.14   Consequential Damages

Consequential damage clauses regarding loss of production time should be included in the purchase order or terms and conditions with the purchase of the machine tool. However, they are rarely agreed upon and less frequently enforced. Enforcement of consequential damages is difficult to prove in a court of law, and it is very difficult to get your company and the machine tool supplier to agree to the uptime of the machine tool.

2.15   Ordering the Machine Tool

One would think that this task would be the purview of the purchasing department. After all, isn't that what they do, purchase? Not in this case. The same team who created the technical specification, gathered the quotes, did the fact-finding, and negotiated the terms and conditions is also responsible for filling out the purchase order and ordering the machine tool.

A purchase of this size will likely require approval from the board of directors and the signature of the president attesting to it. This can take some time. Dare we hope that these directors have read this book, appreciate all the labor and due diligence put into this project, and move forward quickly with the approval process?

2.16   Visit the Supplier's Factory during the Build Process of Your Machine Tool

The purpose of this visit is to check the progress of the build and to make sure the project is on schedule. Get to know who will be appointed as your machine tool supplier's project manager. It is important to visit the machine tool supplier at least two or three times during the build of the machine tool. If you don't constantly check up on the progress of the build of the machine tool, the machine tool supplier will get complacent and not meet the build schedule. At this time you should be able to meet all the necessary people within the machine tool supplier's facility who will be involved in the installation of the machine tool at your facility.

2.17 Acceptance of the Machine Tool at the Supplier's Facility

This is a necessary step in the purchasing process. While your company will incur additional travel costs, this trip should have been included in the travel expense portion of the specification and already approved with everything else. It is far less expensive to travel to the supplier to inspect the machine tool before it ships than to wait until it arrives at your shop, only to discover something isn't satisfactory.

2.18 Summary

By this time, I'm sure I don't have to tell you how serious this business is. There are times it will feel like some kind of wearisome business game, but believe me, it's no game. Nevertheless, always remember that as the purchaser, you are in the driver's seat. Machine tool salespeople know that if they sell your company a product and it performs well, your company is likely to purchase many machine tools from them for years to come. It's up you to make them earn your company's business now and in the future.

At the same time, every project is a new challenge. You can't default to one supplier simply because it performed well on the past purchase. The same due diligence must go into every project as if it were the first. If the machine tool supplier you bought the last one from proves to be the best choice for the next one, that's fine and not at all uncommon. You must put the needs of your company ahead of any other loyalties or friendships.

If a company like yours is purchasing machine tools from the same supplier time after time, it may be because that supplier proves to be the best choice each time. Or, it could be cause for serious concern. Perhaps the project manager and the team are cutting corners on the specification process or even forgoing the process altogether, assuming the next machine from that supplier will be as perfect a fit as the last. Or maybe it's something even worse. Maybe someone somewhere in the decision-making process has an under-the-table kickback arrangement with that supplier. Following closely the process as it is detailed in this book will greatly reduce the risk of mistakes, blunders, and dishonesty.

CHAPTER 3
Layout Phase

3.1   Layout during the Procurement Phase

Since machine tools differ significantly in size and configuration, it will be necessary to consider three or four different locations within the facility. This is the beginning of the layout phase. As the team considers which machine tool to select, an essential consideration is which machine tool(s) will best fit in the physical space that the team will select for placement within the facility. Expect to take great pains in getting this part of the project exactly right. The project leaders will bear responsibility for this and must know the ins and outs of how it's done. Even the most experienced leaders will sometimes engage the services of an architectural and engineering firm to ensure it comes out correct. However, this means that the most detailed knowledge rests with the firm, and the project leader, while certainly still very much in the loop, may not have the exact answers to some of the most complex questions. In my own experience as a project leader, since I'm the one who would have to field questions from all levels, I maintained full control of the layout process.

The layout phase requires developing drawings that show both a plan view and an elevation view. A plan view is simply a drawing of the machine itself as it sets on its foundation when installed and viewed from directly overhead. The plan view is also known as the top view in basic blueprint drawings (see Fig. 3.1). The elevation view is either a right-hand or a left-hand view, or both, detailing the machine tool and the existing foundation. (See Fig. 3.2.) By combining the two views, any trouble spots between the proposed machine tool and the current foundation will become evident. It will reveal how the new machine tool under consideration will work on the existing

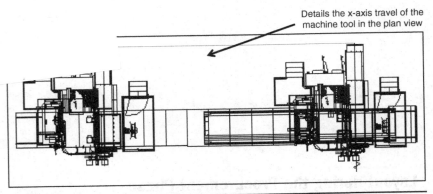

Details the x-axis travel of the machine tool in the plan view

Figure 3.1 Plan view of a machine tool. (*Courtesy of MAG, Giddings and Lewis, Inc.*)

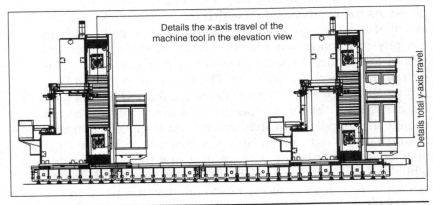

Details the x-axis travel of the machine tool in the elevation view

Details total y-axis travel

Figure 3.2 Elevation view of a machine tool. (*Courtesy of MAG, Giddings and Lewis, Inc.*)

foundation, or if it will work at all. As you interpret the combined views, key considerations are things not included in the drawings, such as the building structure, crane rails, locations of the incoming power source, and other environmental and structural realities. It is also very important to set the travel limits of each moving axis on each machine. This is an essential step in the process of determining the clearance issues between the new machine tool and other machines and structural impediments in the building.

The layout phase must be completed for each machine tool under consideration, and there are usually three or four. This will better enable you develop the budget for the foundation and installation of each.

Another part of the layout phase is the determination of whether the current foundation can be modified to accommodate the new machine tool and ensure its optimum performance. Unless the final foundation is a perfect fit for the machine tool, the machine tool will never operate correctly. It won't machine perfect parts. It will likely damage itself over time and could prove unsafe to operate from the start. However, if the current foundation can be modified to accommodate the new machine tool perfectly, it would save your company a huge sum in addition to capital funding. In the financing of the final plan and machine tool choice, the funds for the foundation are most often included in the price of the machine tool itself and will probably be depreciated at the same rate as the machine tool.

3.2   Maximize Floor Space

One of the most essential yet most ignored tasks in the layout phase is the maximization of floor space. There should be as much open space around the new machine tool as you can possibly carve out of the manufacturing area where the machine tool will ultimately rest. It is inefficient and, far worse, unsafe to have to operate a large and powerful piece of equipment such as a machine tool in a cramped space. Even if the area seems adequate without any modifications or repositioning of things, you could very well be underestimating the area that the machine will require while in full operation. Even if you're absolutely certain that the space you have chosen is adequate to the purpose, the larger it is, and the safer it is. But don't forget to think about the additional floor space that will be needed for the inspection of the workpieces that you plan to produce near the machine tool. I would show at least one or two workpieces around the machine tool in both plan and elevation views. This will give you a glimpse of how much additional floor space might be needed around the machine tool. Therefore please give this process a great deal of consideration during the layout phase.

All machine tool suppliers will provide you with detailed drawings of what is considered a standard installation for their machine tool. At first glance, you might even think it will work perfectly for your particular situation—until you overlay it on a drawing of your current plant layout. Usually, several obstacles will reveal themselves. For example, when you overlay your plant's machine tool layout with the machine tool supplier's drawings, it shows that one of the trenches has to be shortened. The first thing to do at this point is to contact the machine tool supplier and ask if the trench can be shortened without creating safety or operational problems for the new machine tool. However, don't be too surprised if this concept is foreign to them. They sell these things; they don't operate them in all the various environments where they end up.

One plant I worked at simply followed the machine tool supplier's foundation drawings. Either they didn't know anything about the need to maximize the surrounding floor space, or they just assumed that the machine tool manufacturer knew best. In this case, there were several options for repositioning the electrical cabinets close to the wall of the building or within the column lines where there would be no risk of their getting in the way (see Fig. 3.3).

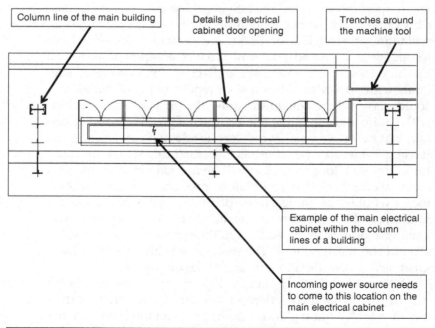

FIGURE 3.3 Electrical cabinet inside the column line.

If it turns out that the machine tool your team thinks is best for your purposes won't work in the designated space unless certain auxiliary equipment is relocated, you can ask the machine tool supplier to handle that for you. This means moving chillers, transformers, coolant tanks, hydrostatic tanks, and auxiliary equipment. The machine tool supplier most likely won't have a problem with that since it wants to sell you a million-dollar or multimillion-dollar machine tool. The supplier won't let a few thousand dollars for a contractor to come in and move things around stand in the way of completing the deal. Take full advantage of this situation, and make the machine tool fit the area in which you want to locate the new machine tool.

3.3   General Notes for Layout of a Horizontal Boring Machine

When you are laying out a horizontal boring machine, it is important to understand the relationship between the foundation and spindle of the machine tool. Most large floor-type HBMs will have a floor plate and/or a rotary table in front of the machine tool. The distance between the centerline of the spindle and the top of both the floor plate and the rotary table should be a minimum of 16 to 20 in depending on whether the HBM has a ram (see Fig. 3.4). This distance is up to the operational department. The *spindle* of the machine tool is also known as the *quill* or *bar*. The ram of an HBM is a large square or rectangular structure, which details support for the spindle. (See Fig. 3.5 for an example of this type of machine tool.)

Keep in mind that the reason that you want 16 to 20 in of distance between the centerline of the spindle and the top of the floor plate is the fact that you don't want the ram to hit the top of the floor plate or rotary table. I have seen machine tool setups where the distance has been less than 4 in from the top of the floor plate. In this configuration, the ram could get into not only the floor plate and rotary table but also into the spindle.

Also, keep in mind that most of the time your workpiece will not set directly on the table or floor plate. There will be machined steel parallels or 1/2/3 or 2/4/6 blocks which you will set between the floor plate and rotary table so that the workpiece will be elevated.

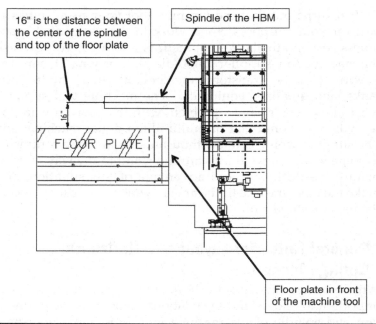

Figure 3.4 Relationship between the spindle for an HBM and the floor plate in elevation view of a machine tool. (*Courtesy of MAG, Giddings and Lewis, Inc.*)

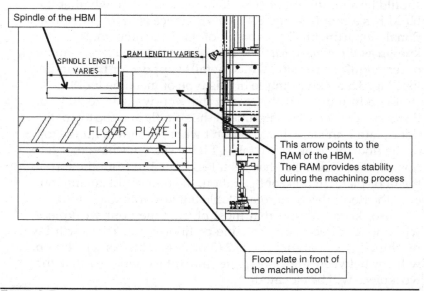

Figure 3.5 Ram type HBM in elevation view. (*Courtesy of MAG, Giddings and Lewis, Inc.*)

Another general layout factor is the location of the chip conveyor and its chip hopper. Based on the purchase of the machine tool, the chip conveyor could be located at many different elevations. You might have more than one chip conveyor or multiple chip conveyors dumping into another. The chip conveyor could need to be tied into the plant chip conveyor system. This will require you to understand what additional electrical controls could be needed.

Maintenance of the chip conveyor must also be considered during the layout phase. Maintenance should have access to the head shaft and tail shaft of the chip conveyor. The location of the main motor and gearbox must also be accessible for maintenance. There should be a section of the chip conveyor where the hinged steel belt can be separated and removed for maintenance. Most chip conveyor manufacturers will sell a watertight construction steel frame for machine tools that requires coolant. These chip conveyor manufacturers will also supply the *chip deflectors,* a bent piece or pieces of sheet metal between the machine tool base and the lip of the chip conveyor (see Fig. 3.6). All the overlapping bolted connections should have silicone to prevent any type of coolant from leaking into the foundation or pits.

One of the major factors in determining the machine tool size will be the maximum size workpiece that you plan to machine. It is a good idea to create a layout in both the plan and

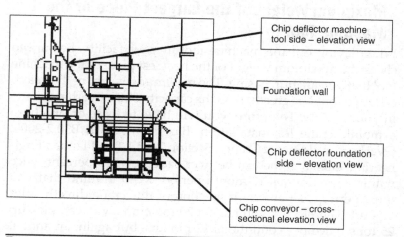

Chip deflector machine tool side – elevation view

Foundation wall

Chip deflector foundation side – elevation view

Chip conveyor – cross-sectional elevation view

FIGURE 3.6 Chip deflectors for a chip conveyor for a machine tool.

elevation views that shows the largest workpiece that you plan to machine on both the floor plate and rotary table of the machine tool. Include all the needed parallels and additional work-holding equipment that might be provided. Make sure to also show the workpiece and the maximum travels of the machine tool. Sometimes you can get the machine tool supplier to provide you with a drawing showing your part on their machine.

Usually, it will be the responsibility of the buyer to install the safety rails and ladders to the proposed machine tool. These items will probably be marked on the machine tool foundation drawings as supplied "by the customer," and therefore the buyer of the machine tool will be responsible for the purchase and installation of both. Most companies like to use their standard safety rails and handrails. Therefore, the buyer should create a drawing detailing the safety rails and ladders needed for safe operation of the machine tool and access to the machine tool foundation and pits. The drawing that details the additional safety rails and ladders will most likely be created, purchased, and installed by the machine tool buyer.

Adding platforms and covers around the machine tool may also be required. It is the responsibility of the buyer to create the necessary platforms and covers to make sure the operators and machinists have safe working conditions around the machine tool at all times.

3.4   Maximum Weight of the Largest Piece of the Machine Tool

When laying out the machine tool in your facility, be sure to check the maximum weight of the heaviest piece of the machine tool that you are purchasing. The maximum weight of the heaviest piece could be greater than the capacity of the overhead crane in your facility. Therefore, you will have to lease or purchase a mobile crane for installation. Based on ASME B30.2-2005, Overhead and Gantry Cranes, Section 2-3.2.1.1(c), Planned Engineering Lifts, a crane can be overloaded by 125 percent twice within a continuous 12-month period. Keep in mind that you must perform an annual inspection of this crane directly after this lift. Therefore if you have a 20-ton crane, you can pick up 25 tons twice per continuous 12 months, but again, an annual inspection of the crane must be performed. There may be one piece of the machine tool that weighs between 100 and 125 percent of the capacity of the crane and therefore can be moved into location and prepped before being set in the permanent location on the foundation. This could save your company hundreds if not thousands of dollars for this one move.

Machine tools also have to be installed where every piece of the machine tool can be put in place with the overhead crane, forklift, or mobile crane. Keep in mind that with an overhead crane, the crane has its travel limits on the trolley, and so you won't be able to install any of the equipment within the column line of the facility or building you are working in. Also understand that some cranes have two hooks on the same trolley. Make sure that you can handle the maximum-weight of the largest piece of the new machine tool or the workpiece that you need to machine with either of these hooks.

It is also a good idea to have the machine tool supplier load the maximum weight allowed on the rotary table of the HBM and perform an accuracy and rapid movement check. This is also the same for a vertical turning lathe (VTL) or horizontal lathe. Have the machine tool supplier install the largest possible weight allowed into the spindle, and check the capabilities of the machine tool's spindle. Experience has taught me that the machine tool supplier doesn't always adjust the spindle or rotary table drives and motors to the largest possible load. This has happened twice in my career and caused much unwanted downtime after the installation of the machine tool had been completed.

If the rotary table of an HBM has a capability of 20 tons, you should require the machine tool supplier to place a 20-ton load on the rotary table at the supplier's facility and run the machine tool in Rapid mode at 120 percent feed rate and check for accuracy and stability issues. This check should also be performed at your facility during operator training or at the end of the machine tool installation.

The same goes for the spindle of a VTL or horizontal lathe. Based on the capacity of the machine tool, place the largest load on it and check the setup of the spindle drive for that machine tool. The main issues that you will run into are the acceleration and deceleration of the spindle drive. The larger the workpiece, the longer it will take to get up to maximum speed and slow down to zero speed. The inertia of the workpiece is something that needs to be taken into account during the setup of the machine tool at the machine tool manufacturer's facility as well as at your facility. Make these checks part of the acceptance test of your machine tool.

3.5 Soil Boring Specification

Because a machine tool is heavy, powerful, and precise, everything about its installation has to be solid and unmovable. The slightest vibration of the machine tool on its foundation or the vibration from some other source will have an expensive negative effect on the machine tool and potentially its workpieces.

No matter how much concrete one pours to hold something up, it's going to be only as stable as the soil upon which it sits. In some parts of the country, the soil is firm and the bedrock is closer to the surface. Of course, this is a good situation. In other locations, such as coastal regions where the soil is moist and sandy, your foundation specification will be markedly different from one used for a firm ground installation.

Some soil conditions may be obvious, but remember, everything about this process has to be precise, and every piece of ground is unique. This is why you must take at least two or three soil borings. You must know *exactly* what is sitting under your machine tool foundation. When you have your soil borings, you will have valuable information that will impact your foundation design significantly.

Create a standard soil bearing specification for each type of machine tool that you will be installing in your facility. The specification should include the following:

1. Number of soil borings

2. Static loading of the machine tool that you plan to insert in the proposed location

3. Dynamic loading of the machine that you plan to put in the proposed location

4. Location of the soil boring based on your facility's column lines (Make sure that you create a drawing for this activity.)

5. Type of machine tool and total weight of the machine tool

6. Maximum weight capacity of the machine tool

7. Maximum dimensions of the machine tool

8. Maximum dimensions of where the workpiece will be located for machining purposes

9. Plan and elevation foundation drawings of the machine tool

10. Plan and elevation drawings of the machine tool on the foundation drawings

11. The spring constant (modulus of subgrade reaction) for the soil bearing to the architectural and engineering (A&E) firm, supplied by the contractor

12. Soil boring data supplied to the A&E firm so that they can design the steel-reinforced concrete foundation for the machine tool

At the completion of the soil boring evaluation, you should give the information to the A&E firm that will design the foundation for your new machine tool. This information will help the A&E firm design the correct type of concrete and size of rebar needed for proper operation of the machine tool.

3.6 Minimum Requirements for the Machine Tool Location

One common blunder in machine tool installation is to decide where the machine tool will go with little or no thought to its surroundings. We've been talking about this a lot in the previous paragraphs, but believe me, there's much more to be said.

A key consideration is the flow of your workpieces through the various stages of the tooling process. Don't just look around the space to decide if it's right for the machine tool. Look *up*. Will the machine tool be where you need it to be in relation to the crane rails? Can the parts for which the machine tool will be used make it to the machine tool without impediment? Can the largest pieces be machined on that machine tool without any overhead obstructions? By now, I'm sure you get the idea: floor space, floor space, and more floor space will be needed in the present and future for your company, and you should think about this process each time you specify, purchase, and install a machine tool.

In addition to space requirements, many other considerations are just as important. The machine tool may be large, heavy, and powerful, but it is a precision device of equipment which doesn't require much to throw it off kilter. It can happen and you may not know it for a very long time. Replacing a machine tool when significantly better technology appears on the market is a good reason to undertake a massive and costly project such as this. Replacing a machine tool because it is damaged beyond practical repair or wears out far too early is simply a calamity. Here are some other considerations that at first may seem minor, but they have everything to do with the performance and life of this investment. At a minimum, the machine tool must be located where it is safe from

- Exposure to direct sunlight
- Significant fluctuations in ambient temperate, e.g., below 5°C (41°F) or above 40°C (104°F)
- Large amounts of dust or corrosive gases in the ambient air
- Rain or other water making contact with the machine
- Ambient humidity of greater than 75 percent for extended periods
- Sudden voltage fluctuations in the internal power
- Excessive electrical noise[1] on the electric circuit powering the device
- Vibration from other machine tools or crane rails, or outside vibrations
- Large overhead door areas where loading and unloading occur
- Having the incoming three-phase power drastically far away from the machine tool
- Having the compressed air and water piping drastically far away from the machine tool

---

[1]An ac power line disturbance is caused by sudden changes in the load. Electrical noise is problematic in solid-state devices because they cannot differentiate between an intended electrical pulse and an unintended electrical spike.

- Constant blowing of air conditioning and heat on the machine tool from the building's climate control systems

Based on my experience, the machine tool supplier and your internal maintenance technicians will request that they turn off the heating, ventilation, and air conditioning (HVAC) systems during the acceptance and accuracy checks of your machine tool. It really does make a significant difference when you are performing laser calibrations of the machine tool.

In machine shops, machine tools themselves will cause considerable electrical noise and fluctuations in the electrical systems of the building. I would strongly recommend purchasing an isolation transformer. This device goes between the power source and the machine tool. It provides an even flow of alternating current (ac) power to the machine tool.

3.7   Develop the Overall Schedule of the Foundation Installation

It is now time to create the first draft of a comprehensive schedule for creating the foundation. We also create the final schedule for the foundation installation in Chap. 4 for the foundation phase, but it is important to develop the first draft of the foundation for your new machine tool in the layout phase to make sure that each phase of the machine tool project can be completed in time. The schedule should cover every task ranging from demolishing the old foundation to completion of the new one. The project leader is responsible for laying out the entire schedule, including providing plenty of time for the new concrete to cure prior to the machine tool installation. This has to be detailed on a Gantt chart program or other scheduling software. This draft schedule of the foundation installation based on the layout views of the new machine tool should include the following items:

1. Removal of the old machine tool
2. Removal of the old foundation
3. Modification of the old foundation
4. Removal of the old utilities (power, air, water, etc.)
5. Installation of the dust walls

6. Receipt of the rebar

7. Concrete pour for the new foundation

8. Completion of foundation

These are the major items that need to be considered in detail during this phase to prepare you and your team members for the foundation phase.

3.8 Planning the Removal of the Existing Equipment

The removal of the existing equipment should be straightforward and simple. In the best-case scenario, your company has already found a buyer for your old machine, and the buyer is going to uninstall it. The project leader must make sure that the old equipment is removed as quickly as possible.

Be prepared to scrap the machine if there's no ready buyer. Many times in my career a company had good intentions of selling a machine tool and ended up just scrapping it. Have the local scrap dealer come in and give an estimate of what he or she would be willing to pay. It may be in the best interests of your company to scrap the machine since it's likely that the used equipment dealers or new buyer will continue to ask questions and request information on the machine tool before deciding whether to buy. The original foundation drawings, mechanical and electrical drawings, or programming manuals will always be things they need, and you probably won't be able to find. It may be a bigger hassle to sell the old machine tool than just to scrap it.

Another thing to consider is that whoever purchases this machine might try to steal the work that you are producing. There are some very good engineers and machinists in the workforce, and they can ask the right questions to figure out how you are machining the workpieces that this old machine was producing. I've had plant managers in the past who would rather scrap the machine than risk this possibility.

Yet another issue that you will likely have to deal with is the cleanup of the old machine tool. The old machine tool's foundation will probably be contaminated with oil and coolant. It will be your responsibility to get the oil and coolant properly cleaned up and disposed of in a timely manner. This will also be costly and must be budgeted for.

3.9   Planning the Demolition of the Old Foundation and/or Floor Slabs around the Old Machine Tool

This is a complex process. It's not simply smashing and removing a bunch of old concrete. When done correctly, this process is often time-consuming and costly. In many cases, it is not necessary. Depending on the condition of the existing foundation and its configuration, perhaps it can be modified to accommodate your new machine tool perfectly. You might also be able to build on top on the existing foundation and elevate the new machine tool and foundation. When this is possible, it saves enormous sums of money and worker-hours.

However, you may not be so lucky, and the old foundation has to come out. The first step is to surround the foundation area with plywood walls. The industry term for these is *dust walls*. As denoted by the name, their main purpose is to contain the massive amounts of concrete dust and dirt generated by excavating the old foundation. Dust walls are good for more than dust containment. They create a safety barrier between shop operations and the foundation demolition/construction project.

They also block the view of the work from the rest of the operation's employees, preventing them from standing around watching what's going on with your project and becoming a distraction to your crew. This is something I learned the hard way. During one foundation removal, I neglected to set up dust walls. I kept getting calls from the facility's safety manager. Each time, it was something as simple as the way we lifted things or how we moved something within the foundation. It seemed odd at the time. As it turns out, rather than attending to their own work, shop employees were devoting their attention to our work. They would look for things they felt we were doing incorrectly and call up to safety officer constantly. There was nothing wrong or unsafe about how the team was conducting the work. It was just that harassing my team was preferable to doing the jobs they were hired to do.

I solved the problem by encircling the entire worksite with 8-ft plywood dust walls. I provided two access points for the contractors to enter and exit the site that were otherwise kept padlocked. Only the contractor's superintendent and I had the keys. This secured the work area and prevented the theft of the contractor's tools and equipment.

3.10   Planning the Removal of the Old Utilities (Water, Power, Air, etc.)

Removing the old utilities is an important part of the preparation of the site. Usually, the internal facilities department handles this task. However, it is not uncommon for an outside electrical contract to come in and do the job if, for no other reason, the internal facilities people simply don't have the available time.

The first step is to locate the entry point of all inbound power to the old machine. Be careful in this step. Most likely, the old foundation has multiple power sources. For example, if the old machine has a pit under its foundation, there is most likely a 115-VAC power feed coming from the plant's general power source and not the machine tool control cabinet. This is common because pits require lighting and electrical outlets for handheld power tools. Aside from the obvious utility of having this capability down in the pit, there is an additional safety factor for the machine tool operator. Let's say, for example, that the machine tool operator accidentally drops a wrench and it falls all the way down to the chip conveyor at the bottom of the pit. To retrieve it safely, the operator must shut down the machine at its power source before descending into the put to get the wrench. If the lighting in the pit is sourced by the machine tool control box, then powering down the machine tool will also leave the pit without lights. This is not a good scenario.

More will be said about this in Chap. 4, Foundation Phase, but for now let's just say that every foundation with a pit needs to have lights, and those lights must be powered by a different source than the machine tool, as the example above illustrates.

As indicated earlier, the machine tool has more than just electricity running to it. Also attached are lines delivering air and water. These, too, have to be carefully disconnected. The conduit for both air and water, as well as the wires for the electricity, should be removed away from the old machine tool and foundation to at least within the column line. At this early stage, you may not know what the power requirements for the new machine will be. It is possible that you could reuse the existing wire and conduit. However, this would be the exception rather than the rule. Usually, it is necessary to remove the wiring and conduit all the way back to the switchgear that fed the old machine tool and foundation.

Another option is to just disconnect the old feeds and leave them where they are and install your new feeds wherever you need them to go. I once worked with an engineer who questioned the need to bother removing the old feed lines when you just may need them again for something else in a couple years. I believe that cleanliness is very important and leaving old, disconnected wires and conduits hanging around is unsightly and doesn't speak well of the project leader. People, usually upper management, will be asking about the presence of the old wiring and conduit for years to come. Avoid that aggravation and just take them down now.

As for the water and air feeds, at this point, it is fair to assume that you don't know the exact water and air requirements for the new machine tool you will soon be installing. However, one thing that I have learned over the years is that you can't ever have enough water and line near a machine tool. For safety reasons, be sure that all stored power is isolated by some locking device such as a power disconnect or something similar. Remember to keep it locked at all times until you are ready to work with it.

3.11   Floor Plate Layout and Installation

When purchasing a large machine tool with large amounts of axis travel in the longitudinal axis, most likely you and your company will require steel floor plate to be purchased and installed in front of the machine tool. Steel floor plate is a large rectangular piece of solid steel, cast iron, or cast steel plate normally with T-slots machined or cast into the width of the plate, allowing the machinist or operator to secure large-angle plates or workpieces. See Fig. 3.7 for an example of steel floor plate

FIGURE 3.7   Floor plate example. (*Courtesy of Challenge Precision Machine, Inc.*)

that you might install in front of a large machine tool. This type of steel floor plate is made out of solid steel.

Also keep in mind that the T-slot dimensions of the floor plate should be confirmed with the machining engineer and operational department for which the machine tool will be purchased and installed. There are many different types of T-slot dimensions, and you don't want to purchase additional T-nuts and tooling for a new type of T-slot if they are not needed.

Note that the T-slots of the floor plate must be laid out in parallel with the longitudinal axis of the machine tool. See Fig. 3.8 for an example. The main reason for this design is the fact that if the T-slots are perpendicular to the machine tool and you start machining and pushing on the workpiece with the spindle and ram of the machine tool, then the angle plates on which the workpiece is setting will most likely push away from the tool.

The specification and the purchase of this type of floor plate are extremely expensive and difficult to justify unless they are purchased with the machine tool. Most large HBM tool suppliers

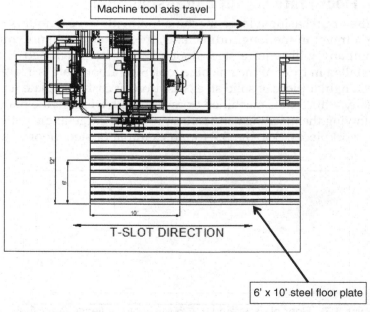

FIGURE 3.8   Floor plate T-slot direction example.

have the capability of purchasing and manufacturing these floor types with their large HBMs. You also have the option to purchase these types of floor plates secondhand from a used machine tool dealer. This option is much cheaper, but you will have to take the responsibility for squaring up the floor plate and installing the floor plate on the new machine tool foundation. Each section of the floor plate should be flat within 0.003 in end to end in a relaxed state. Each section of floor plate should have at least 6 to 12 leveling screws which the installer can level to earth and in relation to the machine tool. Next, it is my recommendation that you use an indicator straight out of the spindle of the machine tool and check the level with the new machine. Of course, this is after the bed of the machine tool has been leveled and aligned and all the axis and laser calibrations have been performed. For safety reasons, I would install the floor plate but only in a rough condition. This allows for the buyer and installer of the machine tool to use the floor plate during the installation of the machine tool safely.

The installation of the floor plate is another issue. Floor plate can be installed in many different ways. See Fig. 3.9 for the first example. This example uses steel shims in certain locations

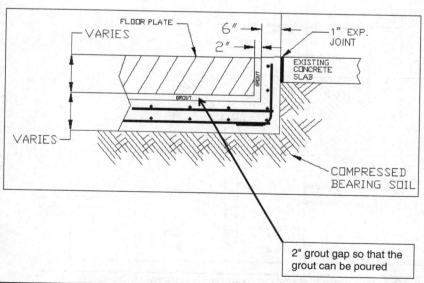

FIGURE 3.9   Floor plate installation example 2 in grout.

under the floor plate and has at least a 2-in gap to pour the grout between the floor plate and concrete foundation. Some floor plate is designed with many cavities and grout pour holes within the floor plate, so that you can pour grout or concrete inside to lock in the floor plate on the foundation. Locking in the floor plate to the foundation provides stability in the machining process. When installed in this fashion, the floor plate can normally be leveled to within 0.125 in from end to end. This is the least accurate way to install floor plate. I would not recommend installing floor plate in front of a machine tool. Normally this type of installation is performed in an area where there is a welding or assembly process that needs the workpiece anchored to the steel floor plate.

Floor plate can also be installed per Fig. 3.10. This will require the general contractor to place or embed steel plate under the leveling screws. The floor plate has anchor holes machined into the plate to anchor the floor plate to the foundation after the leveling has been done. This type of installation is much more accurate than just grouting in the floor plate per Fig. 3.9. Normally with this type of installation, the floor plate can be installed under 0.005 in end to end. This type of installation allows for the floor plate to be checked and releveled as needed.

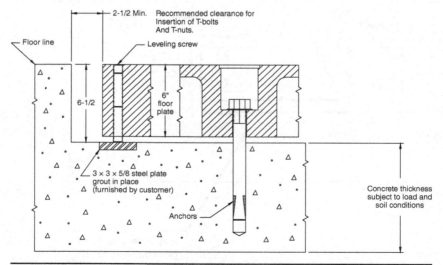

Figure 3.10 Floor plate installation example. (*Courtesy of Challenge Precision Machine, Inc.*)

Another type of floor plate installation uses fixators and/or leveling wedges just like when you install the machine tool. This is the most accurate and expensive option. But it is also the most recommended option for a large machine tool. This option also allows you and your company to continually check and relevel the floor plate as needed. This option will also make your machinist happy since the floor plate is extremely level and she or he won't have to worry about trying to constantly shim or level the part to the machine tool.

3.12 Summary

This chapter on the layout phase is dedicated to understanding the importance of first trying to laying out the new machine tool over the existing foundation. The more you plan in the layout phase, the more prepared you and your company will be during the building of the foundation and the installation of the machine tool. As mentioned earlier, it's possible the existing foundation can accommodate the new machine tool, with or without some modifications. If that turns out to be the case, it will save your company an enormous amount of time and money.

We've discussed the importance of laying out the machine tool in both plan and elevation views. It is also important to lay out the safety rails so that the machine tool supplier and your safety manager will approve the guarding method around the machine.

It is also important to detail the safety rails, ladders, and crane limits on the elevation plan drawings so that both the operations and the maintenance people in your facility understand the limits of loading and assembling the new machine tool.

Sometime during the life of the new machine tool, it will require disassembly for repair or maintenance. It is important to get the operations manager to sign off on the location based on its accessibility for repair and maintenance, as well as the crane limits. In taking this extra step, you will avoid the embarrassing questions that may come a year or two later from a new member of upper management as to who signed off on the location of that machine tool. This documentation could save your job in the future.

CHAPTER 4
Foundation Phase

4.1 Prefoundation Meeting and Foundation Project Leader

The foundation is the block or blocks of concrete designed to support the weight of the machine tool. The foundation is the primary factor in achieving optimum performance by providing the underlaying support-critical machine tools and other equipment. In addition to supporting the machine tool itself, the foundation supports the machine tool with the workpiece being machined, which can be extremely heavy. When you combine the weight of the machine tool, the weight of the workpiece, and the unrelenting vibration created by the machine tool itself while in operation, the strain on the foundation is formidable. Not only does the foundation have to endure all these factors hour after hour, day after day, but also it must do so without ever weakening. It must remain as solid and rigid as the first day it went into operation.*

The vibration of other machine tools, building cranes, plant operations, or other processes must not be allowed to transfer to the foundation. It is essential to structurally isolate this block of concrete from all other sources of vibration. This is illustrated in the installation of a typical floor-type HBM. The only connection between the machine tool and the workpiece is the foundation (see Fig. 4.1).

The commonalities of support-critical machine tools are that they

- Are comprised of several nonconnected segments and large moving masses
- Are designed for producing large workpieces

*Courtesy of Unisorb, Inc.

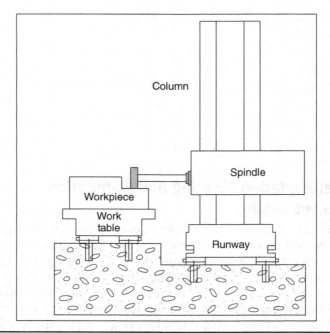

FIGURE 4.1 Support critical machine tool. (*Courtesy of Unisorb, Inc.*)

- Require an anchoring/alignment system, which is provided by the machine tool builder
- Require machine setup procedures that specify in-place alignment criteria
- Require special nonshrink grout in the construction of the foundation

As mentioned earlier, there are a wide variety of machine tool designs and configurations, each created for its own purpose. This book concerns itself solely with machine tools that have multiple pieces and are designed to create heavy workpieces that are moved and loaded with overhead cranes.

Experience has taught me that proper planning, design, and installation of the foundation will require many months of time and work. The cost for the foundation for these types of machine tools is usually equivalent to about 20 to 30 percent of the purchase price of the machine tool. Many factors can affect the final cost of the foundation. The width and depth of the concrete block or blocks that comprise the foundation, including the size of the rebar and the amount of structural steel required, are just the beginning. The number of varying elevations on the top of the foundation and how many trenches the foundation design calls for are also important pieces of the cost equation. Finally, the work schedule required to install the foundation to meet the arrival of the machine tool will also drive the total cost significantly.

4.2   Schedule for the Completed Foundation

It is necessary to prepare the full schedule of the foundation project early in the process of purchasing the machine tool. The limitations on foundation construction within your own facility can drive the buying decision, as well as the machine tool supplier's manufacturing and installation schedule. It's possible that the machine tool supplier will be able to deliver the machine tool even before the foundation is ready. This means a long wait for the supplier with a multimillion-dollar machine sitting in the factory while the foundation is completed. These types of machine tools are manufactured when the purchase order is placed from the buyer. Most times these types of large machine tools are not in stock at the machine tool supplier's manufacturing location. Machine tool manufacturers have an enormous amount of money and worker-hours tied up in that piece of equipment. A long wait to recover such a sizable investment quickly becomes very costly.

Unless the machine tool is being placed in a position where nothing else has ever been installed, the existing equipment and old foundation have to be removed. As mentioned earlier, this can be a rather long and protracted process. If your company plans to sell the old machine to another for further use, then you first have to find a serious buyer. The buyer will undoubtedly want to see it in operation in your facility before committing to the purchase.

Another impediment to the quick removal of the old equipment may be your company's production obligations which require it to run the existing machine until those obligations are met, however long that may take. Once the old machine is finally removed, the lengthy process of demolition of the old foundation and the design and build of the new one now stands in the way of taking delivery of the new machine tool.

Sometimes it's difficult to get the maintenance and facilities departments to make this project a priority and realize that it must be done in a timely fashion. Many expensive problems will result from its delay. One is the overtime required for quick installation of the new foundation or installation of the machine tool. When overtime is required for the general contractors of the foundation and of the machine tool installer, it could pose some serious safety issues. When working a prodigious amount of overtime, employees get tired and worn out and can then become confused and wary. This is the time when people get hurt. So make sure that all team members are performing their tasks to make sure the machine tool's foundation and installation are installed on time.

Another key factor is that after the massive amounts of concrete are poured to make the new foundation, there must be curing time of at least 28 days.

The schedule for the foundation project should include

- Removal of the old machine tool by selling, scraping, or relocating
- Mobilization of the general contractor
- Demolition of the current foundation
- Setting a start date for the new foundation
- Establishing a concrete pour date
- Setting the finish date for the foundation
- Calculating the completion date for the 28-day cure period
- Following the cure, beginning the installation of the nonshrink grout for the leveling devices for the machine tool
- Starting the installation of the machine tool

Once the general contractor has been mobilized, be sure that conditions will not slow or stop his or her work. Most general contractors will issue a change notice against your company and require additional mobilization. This means more time and more money are needlessly wasted.

4.3   Tricks of the Trade (Trench Design and Saving Floor Space)

Please pay particularly close attention to this section of the book. We will be covering some of the most important topics related to the overall project.

First we will discuss how to design a standard trench of a machine tool. Figure 4.2 shows the standard trench design for the machine tool. Of course, the trench is lined with concrete that will be poured at the same time as the main mat for the foundation is poured. The width and depth of the trench can vary from machine tool design to machine tool design. Normally, the trench is situated outside of the main load-bearing portion of the foundation. This is where most designers get the trench design wrong. The angle of both sides of the trench shall have a ½- × ½-in steel bar welded to the top of it. (See Fig. 4.3 for standard angle design for the trench.) The angle will be set in place during the construction of the foundation and captured in place by the concrete pour. This trench design allows for the trench cover to be easily installed and removed during installation and maintenance in the future.

Also, notice in Fig. 4.3 that the steel checkered plate or diamond plate is at least ½ in thick. This is to support forklifts and other heavy moving equipment. The wider the trench. the thicker you should design the steel checkered plate or diamond plate. As part of the plate design, provide for some way to easily remove the steel checkered plate from the trench design. Figure 4.4 shows a standard checkered plate. In the

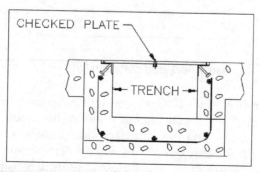

FIGURE 4.2   Standard trench design.

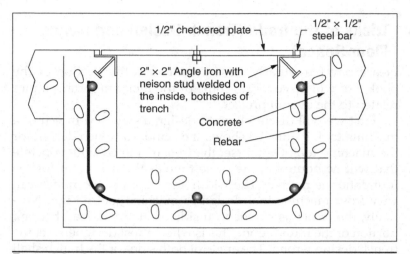

FIGURE 4.3   Standard angle design for trench.

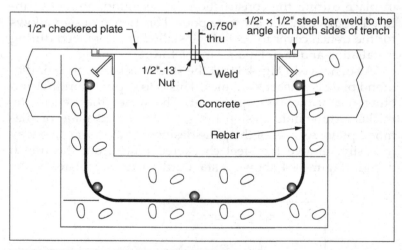

FIGURE 4.4   Standard checkered plate design.

standard trench cover design, you will have to drill a 0.750-in
hole through the steel checkered plate at the center of gravity
and weld a ½- × 13-in nut to the back side of the steel checkered
plate. This provides for easy removal by a crane equipped with
a ½- × 13-in eyebolt or hoist ring. I would recommend no
more than a maximum 4-ft plate in the design. The longer it is,

the larger you should make the through hole and welded nut on the bottom side of the checkered plate.

With regard to covers for the trenches, it is normal for some machine tool suppliers and A&E firms to design bar grating in some trenches. This is a cheaper option than the steel checkered plate or diamond plate but does not have the same loading capacity. I would not recommend using bar grating in a trench if there is any possibility that you could get a forklift in that location. The weight of the forklift will damage the bar grating and will require constant replacement.

However, one advantage that bar grating has over steel checkered plate or diamond plate is that it is easier to remove and modify. But because of the open nature of bar grating, one must take extra care that no trash falls through and accumulates at the bottom of the trench.

Next is the design of a sump in each pit. Some machine tool suppliers will reference the sump design shown in Fig. 4.5. However, this can be extremely dangerous for two reasons: First, when the sump is full of coolant or oil, someone who doesn't understand the sump design could potentially break a leg if he or she stepped off in the uncovered section; second, the sump should be steel lined. This is, in my opinion, just as important, since concrete will crack and the coolant and oil could leak through the foundation. Seam-welding a steel liner inside the sump will save your company plenty in the long run.

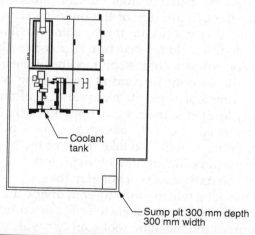

Coolant
tank

Sump pit 300 mm depth
300 mm width

FIGURE 4.5   Standard machine tool sump design.

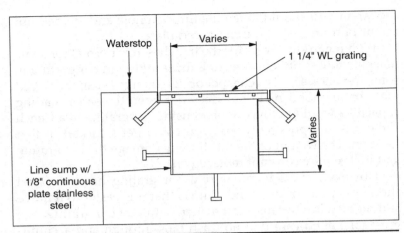

FIGURE 4.6 Recommended machine tool sump design.

See Fig. 4.6 for the recommended sump tank design that should be included in the design of the foundation.

Also, notice the waterstop in Fig. 4.6. The purpose of the waterstop is to not allow oil or coolant to migrate through the concrete foundations at a construction concrete pour. Sometimes these construction pours will be at different elevations during the construction of the foundation, and sometimes they will be close to the sump pit or tank.

Next, install the machine tool control panel. It must be placed within the column line of the building structure. It is also imperative that you locate the machine auxiliary equipment as close to the building structure as possible. By locating the machine control panel in the column line of the building, it will save on the amount of conduit and wiring required to reach the machine tool, since the main services of the building are most likely located in the column lines. Even though it will be difficult to install the main electrical control panel in the column line, doing so will also take it out of the crane travels and space of the building and therefore protect it.

If the main control panel will not fit in the proposed column line, ask the machine tool manufacturer to divide the main control panel into two or three panels to save this additional floor space. Sometimes the machine tool supplier will separate the panels at no cost if the company is anxious to sell a machine.

Even if the machine tool supplier won't do it for free, the added cost will amount to a very small part of the final cost of the machine tool.

4.4 Bury the Conduit and Black Iron Pipe in the New Foundation Only

It is best to bury the conduit in the foundation. This also includes the burying of the black iron pipe for the pneumatic (air) lines. Most machine tools will require some type of coolant. This means that it will need a water source at the machine tool foundation. All foundation designs shall have at least one drawing that details the conduit, water, and air lines running to and within the foundation. This drawing is called the *piping and schedule drawing*.

Most machine tool suppliers will specify in their general arrangement and foundation drawings where to place an air or water connection. However, the machine tool supplier doesn't know your facility and what will be required to run the pipe and conduit over to those connections. Therefore, the A&E firm that helps you design the foundation should detail the placement of the pipe and conduit to supply air, water, and power and how all this is to be buried in the foundation.

When factories are built with the intention of housing many machine tools or pieces of equipment that require large foundations, burying any conduit outside the machine tool foundation should be avoided unless it is close to the pile caps of the building and next to a trench that enters the foundation. In other words, do not install the conduit and piping under the standard concrete slab in the middle of the factory if you have any idea or possible future plans that some other machine or piece of equipment will be installed at that location at some time. Always place every electrical service in a place where it is easily accessible. When it is properly done, all electrical connections, conduit, power tracks, junction boxes, and pull boxes are always installed overhead and in the building structure of the factory. This also allows you to isolate, check, and service them as needed.

Do not to run any conduit under the concrete slab next to the building pile caps and footings. While you are designing the machine tool foundation, always have one of the trenches installed next to the building pile cap or footing. This will make it extremely easy to add any new service or connection to the machine tool from the building column line and structure.

4.5 Foundation Pit Modifications

There are many foundation modifications to the standard machine tool foundation that should be considered. First, if a foundation is required for a machine tool, it should be recessed approximately 4 to 6 in from the factory floor elevation. See Fig. 4.7 for an example of this improvement. The purpose of

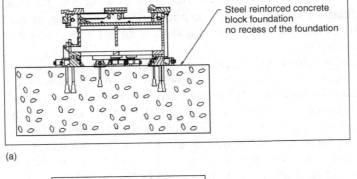

Steel reinforced concrete block foundation no recess of the foundation

(a)

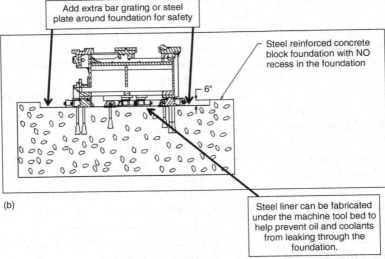

Add extra bar grating or steel plate around foundation for safety

Steel reinforced concrete block foundation with NO recess in the foundation

6"

(b)

Steel liner can be fabricated under the machine tool bed to help prevent oil and coolants from leaking through the foundation.

FIGURE 4.7 Recommended machine tool foundation design with recess in foundation.

this recess is to contain the coolants and oils that are needed to operate the machine tool. These lubricants are not easily contained. The recess will help you keep them isolated within the machine tool foundation. I also recommend that the recess be coated with two-part epoxy paint or lined with steel. If you choose a steel liner, it must be seam-welded and watertight. Lubricants must never leak into the surrounding soils, as this would be an environmental disaster for your company. The A&E firm can easily include this in the foundation design, and the general contractor can install the steel liner during the installation of the foundation. Of course, this is will be more expensive than the two-part epoxy paint, so your company must be willing to pay the additional cost.

Every pit and trench around the foundation should been sloped to a low point where a sump pit or tank and the coolant pit and tank can be installed to collect the liquids and filter and reuse the liquids, if possible. The sump or large coolant pit or tank must be steel lined just as in Fig. 4.6. The steel-lined tank is not expensive to build and install because the general contractor will have to build temporary forms anyway during the build of the foundation if he or she doesn't elect to install a steel liner. If the slope of the foundation is installed incorrectly, this can cause extreme issues. They are detailed below.

1. Coolant will not flow back to the coolant pit or tank.
2. Oils will not flow back to the sump pit or tank.
3. The coolant tank must be constantly refilled with coolant and water.
4. There must be constant removal of the contaminated coolants and oils.
5. Modifications to the foundation pits and trenches to get the coolant to flow in the proper direction will be expensive.
6. There will be constant evaporation of the coolant/water because the slope of the foundation is not correct.

Another design feature for every pit is the means to access it. Every pit should have a ladder or stairs for entry for operational and maintenance issues. You should ask the EHS manager of your facility to review the ladder or stairs, as well as the entire machine tool pit configuration *before* the design and build of the foundation; including purchasing the ladders or stairs. Most likely, the machine tool supplier won't supply the ladder. It will then be the responsibility of the general contractor or machine tool installer. Preferably, the general contractor will be responsible for the purchase and installation of the ladder, as she or he will need to access the pit during the construction of the foundation.

If the machine tool uses coolant through the spindle or flood coolant, most machine tool manufacturers will include a coolant pit design in the foundation. Machine tool suppliers will furnish you and your company with a general arrangement or foundation drawing showing the needed location of the coolant pit. The *coolant tank* is the reservoir that holds the coolant required for the machine tool. It is part of the machine tool and will likely be included in the purchase of the machine tool. The coolant tank will have all the needed pumps, filters, controls, valves, and level switches needed to supply sufficient coolant for proper operation of the machine tool.

The *coolant pit* is a reservoir that is part of the machine tool *foundation*. The coolant tank will receive coolant from the coolant pit. The concrete foundation of the machine tool should be sloped gradually toward the coolant pit. Just like the sump pit, the coolant pit should be steel lined and inexpensive for the general contractor to build and install. The main difference between the sump pit and coolant pit is that the coolant pit will be much larger.

Correctly sizing the coolant pit is important. The coolant pit should be 2 to 3 times the size of the coolant tank. As the machine tool draws coolant from the coolant tank, the coolant is supplied to the spindle either from the face or through the spindle. Coolant sprays on the cutting tool. It will cover the cutting tool, workpiece, and therefore all the machine tool foundation. The coolant will then gradually make its way back to the coolant pit through the slope of the foundation. The greater the foundation slope, the faster coolant returns to the pit. The better the foundation slope, the faster the coolant will return to the coolant pit. It is imperative that the coolant pit be 2 to 3 times the size of the coolant tank to allow time to for the coolant to make it back to the pit so the tank does not run out of coolant, cause the machine tool to alarm, and shut down the machining process.

It is also important to clear away the metal chips continually to allow the coolant to flow better on the foundation. Metal chips must be cleaned from the foundation at least daily. Keep in mind that the coolant will also travel through the chip conveyor. For this reason, it is imperative that the chip conveyor you purchase be watertight. If it isn't, the conveyor could leak coolant into places where the foundation was not designed to be sloped. In my experience, a majority of the coolant will travel through the chip conveyor and back to the coolant tank or coolant pit. Therefore it is important that the chip conveyor be watertight and slope gradually toward the coolant tank or pit. Again the faster the coolant travels back to the coolant tank and coolant pit, the fewer issues you will have in the operation of the machine tool in dealing with the coolant system.

4.6   Develop a Foundation Design Specification

The *foundation design specification* is the document that details the layout and design of the required steel-reinforced concrete foundation for the machine tool your company has purchased. Usually, the design and build of the foundation will be the responsibility of the buyer. Machine tool suppliers are usually willing to do this for a machine tool they sold, but it is likely to be very expensive. It also can create real problems. The machine tool supplier probably isn't familiar with your facility is unlikely she or he will know much about the soil conditions at your site. The additional cost of travel for the machine tool supplier and the added unknowns that the supplier will encounter will drive the cost of the foundation design upward. Rather than handle the task themselves, machine tool suppliers will likely engage a local A&E firm to help design your machine tool foundation; this is another added expense. Thus it is better for your company to tackle this task.

The following is a list of standard items included in the foundation design specification.

1. Structural evaluation of existing foundation, if needed

2. Finite element analysis of the foundation to evaluate deflection, if needed

3. Demolition/shoring plans and details of the new foundation

4. New machine tool steel-reinforced concrete foundations plans and details in AutoCAD DWG or DXF form

5. Grout pocket and anchor support plan and details

6. Metal chip conveyor trenches and platform framing plans around the machine tool

7. General notes and specifications

8. A complete set of "as built" project drawings created electronically at the project's conclusion

9. Concrete material specifications that appear on the drawings

10. Piping and schedule drawing for the new foundation

11. One plan and elevation drawing that details the new machine tool on the new foundation

12. Five foundation design reviews as discussed in Sec. 4.7

The structural evaluation of the foundation and the finite element analysis to evaluate deflection are not required for all foundation designs. With the very real possibility that the existing foundation can't be used, the time and cost involved in performing a finite element analysis for the foundation will be expensive. Usually, the foundation can be designed without these first two items. A finite element analysis should be performed only when you have issues with the soil conditions where you plan to place the machine tool. If the intended machining process is critically exact, this type of evaluation is necessary.

It is extremely important that you receive the drawings of the foundation design in computer-aided design (CAD) format. As a minimum, you should request these drawings in AutoCAD form, which is .DWG or .DXF form. With today's technology, it is acceptable to ask the A&E firm to develop three-dimensional CAD drawings of the foundation. Modern three-dimensional software allows the designers to easily perform a finite analysis of the foundation if needed. The three-dimensional drawings also aid the manufacturing engineers of your company to lay out the machine tool for its intended machining processes.

4.7 At Least Five Review Sessions for the Foundation Design

The machine tool foundation will be a permanent structure at your facility for many years. Therefore, it is important to have the foundation well documented, not only for your own reference far into the future, but also for the facility engineers and operational managers to reference after you have retired or left the company. This is one reason why at least five design review sessions of the foundation design, with the A&E firm, are needed. Another reason is that going through these five reviews forces the whole team to examine thoroughly all the issues associated with the foundation design and its relationship to the machine tool. This, in turn, will guarantee that the foundation will be functional for the machine tool. These five design reviews with the A&E firm are as follows:

- The first design review is called *Issue for Comment*. This is the first review of the foundation with the A&E firm, facilities department, and project team. It's important that you forward this first set of foundation drawings to the machine tool supplier for review and comment. Be sure to double-check that all the notes on the machine tool supplier's foundation drawings and the general arrangement drawings were added to the foundation construction drawings provided by the A&E firm. One example is the painting of the foundation with epoxy paint after the concrete and the nonshrink grout have cured. This is standard on nearly every set of machine tool foundation drawings. Expect the concrete foundation to have some cracking, which is why nonshrink grout and painting the foundation are necessary. All machine tools have some sort of lubrication oils and coolants that leak onto the foundation floor. If not taken into account in the planning phase of the foundation design, these will cause major environmental issues. Your company's EHS director will be ever grateful that this eventuality was considered early in the process!

- The second design review is called *Issue for Revision*. This review will determine whether some changes are needed to the first design review. Inevitably, there will be. At this point, the team should list every item requiring modification and provide that list to the A&E firm. Be sure that the foundation is sloped properly and that every elevation is correct per the machine tool foundation and general arrangement drawings when compared to the A&E firm's drawings.

- The third design review is another Issue for Revision step. In this step, the team should confirm that there is sufficient clearance for every moving part of the machine tool between it and the foundation. Equally important is to make certain that the machine tool doesn't interfere with the crane or the building columns located in and around the machine tool.

- The fourth design review is called *Issue for Approval.* This is the final opportunity for the team to make changes. You should know in advance that most A&E firms won't like being asked for five design reviews. At this point, the team may decide that there no additional changes are needed, and a fifth design review isn't necessary. This is a good time to send the design to the general contractors for bid. The earlier the general contractors start reviewing the foundation drawings, the better understanding they will have in quoting and building the foundation.

- The fifth design review is called *Issue for Construction.* At this final juncture, the foundation drawings are ready to go out for bid to at least three or four general contractors. This is also when the engineering firm will "stamp" or "seal" the drawings. Make sure to get at least two E-sized copies of sealed or stamped drawings from the A&E firm. Depending on the state and local governments, the build of the foundation may require building permits. This makes it imperative to get the final sealed and stamped drawings from the A&E firm.

Once the general contractor is selected, she or he is bound to have issues with the foundation's design. The general contractor's people have to build it. It's not uncommon for a general contractor say there is no possible way to build this foundation. General contractors usually ask to change at least one or two items in the foundation design. This is why you must reserve the right with the A&E firm to have at least one more design change based on the general contractor's constructability concerns.

During the build of the foundation, the general contractor must have at least one working set of foundation drawings located on the new foundation at all times. Each change to the foundation should be "red-lined" and signed off on by the project manager or someone else on your project team. Red-lining is the procedure for marking the change on the working copies of the foundation drawings with a red pencil or pen, then signing off on it with your signature and the signature of the superintendent of the general contractor. It is also important for the project manager to have a duplicate set of foundation drawings with the same red-lined changes. By the end of the foundation build, the working drawings are usually in such bad condition that you will need to use the duplicate set for the "as built" drawings.

The as built drawings are returned to the A&E firm for final detailing and documentation. After the A&E firm updates the as built drawings with its CAD software package, the updated files are sent back to the project manager of the build of the foundation and the project team. This is the final documented step in the foundation drawing phase.

4.8 Bid Foundation Design and Purchase Order for the Foundation Design

As with the purchase of the machine tool, it is important to have at least three A&E firms bid on the design of the foundation. It is also important to apply the same terms and conditions placed on the purchase of the machine tool to the design of the foundation. The cost of the foundation design usually amounts to less than 5 percent of the total cost of the foundation. Once the A&E firm is selected, prepare the purchase order for the design.

4.9 Installation of the Dust Walls and Limiting Access during Construction

Dust walls serve two purposes. The first purpose is to completely surround and barricade the foundation from the rest of the existing manufacturing equipment and processes during the

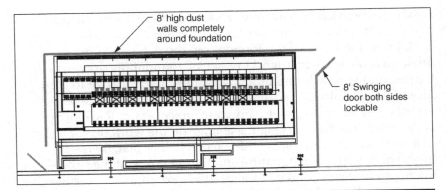

8' high dust
walls completely
around foundation

8' Swinging
door both sides
lockable

Figure 4.8   Dust Walls around the Machine Tool and Foundation.

demolition of the old foundation and building of the new one. The dust walls should be at least 8 ft high and should leave at least 4 to 6 ft between the dust walls and the edge of the foundation demolition. This is done so that the necessary personnel can move easily and safely around the build of the machine tool foundation. The second purpose for the dust walls is to detour sightseeing. Depending on the size of the machine tool that you have purchased and the size of the foundation, it will probably take many months to complete. The 8-ft dust walls prevent the other employees from stopping every day to watch the general contractor perform the work of building the foundation. This alone will prevent a constant barrage of useless questions. At the end of the workweek, the general contractor can secure the entire foundation by bolting down and locking the dust walls from both ends. Figure 4.8 illustrates the type of dust walls described in this step.

The dust walls should remain in place until the machine tool safety rails protecting the open foundation pits are installed. Also, it might be a good idea to go ahead and leave the dust walls up during the build and installation of the machine tool.

4.10   Purchase Order for the Demolition of the Old Foundation and Installation of the New Foundation

This is purchase order is second in capital funding only to the purchase of the machine tool itself. The purchase of the foundation is just as important as the purchase of the machine tool. Often, the foundation design and installation is viewed as too costly when added to the cost of the machine tool. Plant managers and controllers often ask why the foundation requires so much money and time. The main step in this process is to allow the purchasing department to facilitate the purchase of the foundation.

The project manager's role in the foundation purchase is straight-forward. The foundation will be installed per the sealed and stamped engineering drawings. The terms and conditions of the purchase order for the foundation won't differ much from the purchase of the machine tool, and purchasing department is allowed to perform its function in protecting the commercial interests of the company.

4.11   Review and Sign-off on the Rebar Drawings for the New Foundation

During the build of the foundation, you should obtain the rebar drawings from the general contractor for your review and approval. By asking for the rebar drawings, you hold the general contractor accountable for getting the rebar in time. By asking for and double-checking these drawings, you force the general contractor to be on time with the rebar based on the foundation installation schedule. It also confirms that the rebar ordered by the general contractor matches the needed rebar for the machine tool. You don't want the general contractor trying to piece together the rebar of the foundation within a rebar drawing.

4.12   Prejob Safety Briefing prior to Demolition and the Installation of the New Foundation

The prejob safety briefing is essential before the start of the demolition of the existing foundation and the start of the new foundation. The most important part of this process is to make sure nobody is seriously injured or killed during the demolition of the old foundation and installation of the new one. This may sound extreme, but a severe injury or a death at the build site is a real possibility. No one wants it to become a reality. Review the personal protective equipment (PPE) rules and safety program of your company with each contractor. Implement some type of training sign-off sheet for this activity. It's advisable to perform a safety review of the PPE rules at least once per month. PPE includes safety glasses, hard hats, hearing protection, face shields, rigging equipment, and anything else associated with the safe build of the foundation.

It is my opinion that hard hats should be used throughout the entire build and installation of the machine tool foundation. This makes it easy for you, the project manager, and the project team to easily count the number of general contractor employees who are working on your foundation on a daily basis.

4.13   Concrete Cylinder Test

The A&E firm should specify a concrete cylinder test in the details of the foundation drawings. This test involves using the same concrete used in the new foundation to make hollow compression cylinders, which are then tested under intense pressure to find their breaking point. This measures the compressive strength of the concrete after 7 days of curing and again after 28 days. Give the standard 7-day and 28-day test results to your maintenance and facilities department.

4.14   Foundation Cure Period

Based on most concrete studies, concrete reaches 99 percent of its compressive strength at 28 days. This time must be factored into the overall schedule of the foundation and machine tool installations. Keep in mind that you can start the installation of some parts of the machine tool before the 28-day curing period has ended. For example, you can install the electrical control panel and start the installation of the electrical service to the machine tool much earlier than 28 days. The chip conveyor can be installed during the 28-day curing period. The chip conveyor has almost no load bearing on the foundation. It should be one of the first pieces of equipment installed. I would recommend that no machine tool equipment be installed on the foundation before the 7 days of concrete cure.

4.15   General Contractor Specification

It's probably a good idea to create a general contractor specification that details the requirements which a general contractor will abide by while working at your facility. This is a specification that your team and other departments will be able to reference when bringing in all types of contractors to work at your factory. Below is a list of items that should be included in the general contractor's specifications:

1. Safety requirements	Detail the minimum safety requirements for general contractors working at your facility. Safety glasses, steel-toe safety shoes, hard hats, and ear protection are some examples of items you would force the general contractor to use at your factory.
2. General requirements	The general requirements detail the fact that the contractor will keep the area neat and clean during the construction, the length of the project, and the construction required; and they let the contractor know that the buyer will continue its normal manufacturing operations during this construction. Thus sometimes the general contractor might have to wait to use the crane or forklift if it is needed by manufacturing or maintenance. Also, specify that no gambling, drugs, firearms, weapons, or fighting will be tolerated at the buyer's facility.
3. Hazardous materials	The general contractor is not allowed to bring any type of hazardous materials onto the premises of the facility unless it is approved by the environmental, health, and safety manager. Also, any type of hazardous material brought to the site shall have its Material Safety Data Sheet (MSDS) reviewed by the EHS manager.
4. Overtime	All overtime associated with the general contractor and on this project will be the responsibility the general contractor and not included in the cost of this project.
5. Working hours	This topic specifies the working hours for the general contractor at your facility. This item is extremely important. The general contractor will take advantage of the working hours if you don't specify them up front. For example, if their workers come in early one or two days a week, it will force you to come in—or communicate to the other nighttime or evening supervisors—to watch them. Sometimes this creates more issues and harm than just keeping the general contractor on a consistent schedule.

6. Shipping and receiving of general contractors materials, tooling, and parts.	Specify the physical address for the general contractor to use to ship materials, tooling, and parts at the facility. Also, detail the overhead door and area in the facility where the items are to be delivered. Let the general contractor know the name of the project manager or team member associated with all the general contractor's materials, tooling, and parts.
7. Storage of tooling and materials on site	Put in writing that the general contractor will be allowed to store tooling and materials on the buyer's facility, but all tooling and materials should be stored and locked by the general contractor. The general contractor assumes the risk of damages associated with the storage of materials at the buyer's facility.
8. Utilities	Detail in the specification what utilities the buyer will allow the general contractor to use at the buyer's facility, e.g., compressed air, water, argon for welding, three-phase power, single-phase power, and nonpotable water.
9. Hot work permits	Detail in the specification what type of process requires a hot work permit, e.g., welding, brazing, grinding, and soldering. You really don't want the general contractor performing this type of hot work in your manufacturing facility without having the general contractor fill out the proper hot work permit and let you or the maintenance department review and sign it for a limited amount of time.
10. Use of overhead cranes and forklifts	Detail in the specification whether the buyer will allow the general contractor to operate the overhead or jib cranes in the facility where the general contractor is performing work. Also, detail in the specification whether the buyer will allow the general contractor to use the buyer's forklifts or if the general contractor should bring her or his own forklift.

This general contractor specification is something that can be specific to just one department, if needed. For example, the maintenance and facility department would have a standard general contractor specification for its type of repair and service work, and the operations department would have a standard contractor specification for its manufacturing type of work to be performed.

The general contractor specification is a document that the purchasing and safety departments can reference for the

1. The design and the construction of this foundation are the customer's responsibility. The purpose of this foundation drawing is to indicate proper location, holding, and leveling of the machine and auxiliary equipment on the foundation. The foundation is to be designed and built to suit soil conditions at its location and to withstand the machine weight and loads as indicated within the deflection range outlined in note 2.

FIGURE **4.9**   Design and Construction of the foundation is the customers responsibility. (*Courtesy of MTR, Inc.*)

machine tool supplier, the general contractor who builds the machine tool foundation, the machine tool installation company, or even the electrical installation company. I would also put the general contractor specification in the purchase order for the machine tool and the build of the foundation. Hopefully these items will cover you and your company in the long run.

4.16   Foundation Summary

After the foundation is installed and the 28-day cure period is completed, it is advisable to place a heavy piece of equipment or a block of concrete on the new foundation to see if the foundation has moved or will move. If you have time to do this before the installation of the machine tool, you should. It will provide a real-life indication of whether the engineering design firm did a good job in building the foundation. The general contractor, using optical transits and leveling equipment, can also perform this check easily.

From the foundation design specification to the foundation bidding process, to the building and completion of the foundation, the project manager or engineer is responsible for everything. Again the design and the construction of the machine tool foundation will be the responsibility of the machine tool buyer. Most machine tool suppliers will have some sort of note on their general arrangement and foundation drawings detailing this issue. See Fig. 4.9 for this type of note. The steps I have outlined in this chapter will give you a plan for tackling these tasks.

CHAPTER 5

Installation Phase

5.1 Create the Machine Tool Installation Specification

The *Machine Tool Installation Specification* is a document that details the agreed upon items to install the machine tool to OEM standards by the buyer or an independent installation company. The best option is to have the machine tool installation created by the machine tool supplier and/or your own internal maintenance and facilities group. If this is not done, then the project manager will have to create this machine tool installation specification. During the purchase of the machine tool, make sure the machine tool supplier provides a mechanical and electrical installer during the installation.

Usually, the buyer of the machine tool also has a maintenance and facilities department that helps with the installation of the machine tool. This department has mechanics who can run the cranes, help with the leveling of the machine tool, install the headstock, and fill the correct oils, among other things. The electricians can also help run the crane, wiring up the power track to the main control panel and connecting all the auxiliary devices.

This type of installation training is extremely invaluable for understanding how the machine tool is built and how it works. All mechanics and electricians working on the installation should go through the machine tool operational and maintenance training, which is conducted at the machine tool builder's facility and at the buyer's facility.

Below is a list of standard items and some details that should be included in the machine tool installation specification.

1. *Details of the type of machine tool to be installed.* Put the machine tool supplier's contact information in the

specification, and let the supplier know who will be quoting the installation of their machine tool in your facility.

2. Purchase and installation of the grout cans, grout forms, and nonshrink grout.

3. *Installation of the machine tool oils and lubrication.* The purchase of the machine tool oils and lubrication is the responsibility of the machine tool buyer. The machine tool installer is responsible for providing the pumps and hoses needed to fill the tanks and reservoirs, including the first fill of the coolant needed for the machine tool.

4. *Crane and forklift operation for the machine tool.* The operation of the crane can be a real distraction during the machine tool installation. The installer and the manufacturer want the use of the crane all the time. They must understand that your company has a business to run and that the availability of the crane is in the hands of the operations department. Therefore, the operation of the crane is also something that should be detailed in the machine tool installation specification. The same is true for the use of the machine tool buyer's forklift, if it to be used. It is best if the machine tool installer provides for a dedicated forklift in the quote. I would also recommend that the machine tool installer provide the needed scissor lift, worker lift, or any other type of installation equipment if needed.

5. *Installation of the checkered plate and bar grating around the machine tool.* This will be one of the largest costs and most complex topics of the machine tool installation. Drawings detailing all the checkered plate and bar grating around the machine tool will be necessary. The drawings are given to the machine tool buyer for review and sign-off. These drawings are also prepared in AutoCAD format for later review.

6. Special lifting equipment that may be needed for installation of the machine tool.

7. Manufacture and installation of the baffle or chip deflectors between the machine tool and the main chip conveyor for the machine tool.

8. *Purchase and installation of the safety rails around the machine tool.* The machine tool installer will supply

detailed drawings of the safety rails for review and sign-off by the machine tool buyer.

9. *The machine tool installer provides all labor to properly install the machine tool.* This includes all welders, machinists, electricians, millwrights, pipe fitters, and fabricators needed to complete the installation.

10. *The machine tool installer provides all materials required to properly install the machine tool.* This includes tools, wrenches, conduit, power tracks, tools, pipe, wire, and the special alignment tools needed to complete the work. Any special tooling that is required for operation of the machine tool shall be detailed and a list developed for the machine tool buyer.

11. *The machine tool installation specification includes all travel and expenses for all labor required to install the machine tool.* Regardless of the number of trips to and from the machine tool supplier's facility and the machine tool buyer's facility, all expenses should be covered in this specification.

12. The machine tool installation specification must include all ladders and stairs needed to access the machine tool foundation pits.

13. *The machine tool installation specification should specify whether it is a fixed-price contract or a time and materials (T&M) installation.* In my opinion, a fixed-price contract is much easier to manage but is most likely more expensive than a time and materials contract. A fixed-price contract with an installation schedule attached to the purchase order of the machine tool installation almost guarantees that the machine tool installation company won't drag its feet during the installation process.

14. Terms for payment of the machine tool installer.

15. *Liquidated Damages.* The parties must agree upon a predetermined sum to be paid to the machine tool buyer in the event that the machine tool installer fails to install the machine tool in a timely fashion. Normally, this is 5 to 10 percent of the cost of the installation of the machine tool. Most companies won't go for more than 10 percent.

16. *Work schedule for the machine tool installer.* Usually, the machine tool installer will want to work a minimum of

10 h/day, 6 days/wk. This may be problematic, as the crane will be in use by the buyer's personnel exclusively. If you allow the machine tool installer to work 6 days/wk, then you will have to pay your crane operator overtime to support the machine tool installer. Another factor is the work schedule of the machine tool supplier. Whatever schedules the contract states for the machine tool supplier should be the same for the machine tool installer or your in-house maintenance and facilities department. The work schedule should also state whether the machine tool installer will work weekends and holidays if needed. The installer must also let the machine tool buyer know of any planned work stoppages during the installation, at least 2 to 3 days in advance.

17. *Workspace and telephone access for the machine tool installer.* The machine tool buyer should allocate to the installer a quiet workspace with computer and telephone access, to call her or his company if a problem arises. If you don't allow the installer this type of access, most likely the project manager or another team member will constantly be asked to be a go-between for the person installing the machine tool and the machine tool installation company. This will take up a lot of unnecessary time for you, the buyer of the machine tool.

18. *Items that the machine tool buyer is responsible for.* The machine tool buyer will provide wood or other cribbing for the setting of machine tool parts, as well as adequate lay-down space to stage machine tool parts.

19. Adequate three-phase and single-phase power for installation of the machine tool.

20. *All applicable documents supplied by the machine tool manufacturer.* These documents include foundation drawings, lists of oils required by the machine tool manufacturer, and details of the nonshrink grout.

21. *Measurements of the as-built foundation in relation to the proposed foundation drawings.* It is the responsibility of the machine tool installer to double-check the as-built drawings and measure the as-built foundation for any issues related to the build of the machine tool foundation. I recall one instance that serves as a

Department	Team Member	Sign-offs for the Draft Machine Tool Installation Specification
Operations/ manufacturing	Manufacturing engineer Machinists	
Purchasing	Capital purchasing agent Purchasing agent	
Facilities	Electrical facilities engineer Mechanical facilities engineer Plant engineer	
Maintenance	Controls engineer Mechanic Electrician	
Accounting	Accountant	
Engineering	Machining engineer	
Programming	NC programmer	
Quality	Quality engineer	

TABLE 5.1 Machine Tool Installation Specification Team and Sign-offs

good example: A general contractor forgot to drill a kicker hole in the foundation. This required the general contractor to go back and core-drill the kicker hole *during* the installation of the machine tool.

After the completion of the machine tool installation specification, create a sign-off procedure so that other people within the organization can review and approve the specification. Table 5.1 shows a list of team members who should review the machine tool installation specification.

5.2 Installation of the Machine Tool without Maintenance and Facility Support

If the machine tool supplier performs the machine tool installation, your own internal maintenance and facilities group should assist. There are suppliers in the industry that will install the machine tool without any help from your people, but if you choose that route, you will forfeit the valuable knowledge the machine tool supplier can pass along, including the first alignments and calibration. This information helps in developing the preventive maintenance checks and in maintaining and tracking the alignment and calibration data for years to come. Hence, there are machine tool installation companies that serve as contract machine tool installers and contract maintenance and facilities providers. These tasks differ greatly. Some companies will be proficient in one and not in the other.

The contract machine tool installation companies will install the machine tool, help perform the alignments and calibration, and help start up the machine tool. These companies can be simple fabrication, construction, and/or electrical companies. The machine tool supplier, not the installation contractor, bears ultimate responsibility for setting up the machine tool and completing all the alignments and calibrations per OEM requirements. The supplier will oversee the technical piece of the installation to ensure that all settings and calibrations comply with strict OEM requirements.

After the installation of the machine tool, the contract maintenance and facilities company will also maintain the machine tool for a contracted amount of time. The problem with these companies is that they usually experience constant turnover of personnel. The internal maintenance and facilities departments of most machine shops are usually quite stable and enjoy a high level of employee retention. Retaining valuable employees means retaining valuable knowledge of each and every machine tool at your facility. If you work for a company that is thinking about going to a contract maintenance and facility company, I would reference this section of the book and try to defer discussion of this topic with upper management as long as possible.

During the installation, it is important to get at least two or three drawings of the completed machine tool. Whoever works with the machine tool supplier will need these drawings for installation.

5.3   Machine Tool Installation Bid—Get at Least Three Quotes

The project leader and your company's purchasing agent must be on the same page, especially for this step. The installation of the machine tool is just as important as the purchase and build of the foundation. The machine tool installers deserve the same respect as the machine tool supplier they are working with.

The machine tool installation company bids the installation based on the machine tool installation specification. It is a good idea to bring all the prospective bidders together in a prebid specification meeting. Hand out the machine tool installation specification, and go over it item by item. The goal is to make clear the scope of work required for this specification. It is also good to have the machine tool supplier attend the meeting to answer any questions that arise regarding the machine tool installation specification. Limit this to *one* meeting so that the potential machine tool installation bidders all have a clear understanding of what they are being asked to quote, to ensure an "apples to apples" bidding process. It is also a good idea to let the machine tool supplier quote on the entire installation of the machine tool. Most likely the machine tool supplier will provide the highest bid for the installation of the machine tool, but sometimes it can be the lowest. In my opinion, if your company can afford it, I would recommend that the machine tool supplier also perform the installation of the machine tool. Most of the time this doesn't happen, and it will definitely create much more work for the project manager and the team members.

5.4   Machine Tool Installation Purchase Order

The machine tool installation purchase order will represent what is likely the third-highest cost associated with the machine tool. The installation cost of the machine tool can be as high as 10 or 15 percent of the cost of the machine tool, which is why it is important to place the same terms and conditions on it as were placed on the purchase of the machine tool, the design stage of the foundation, and the build stage of the foundation.

After receiving all the quotes from the machine tool installation companies, you and your purchasing agent should meet and decide which installer will be the best option based on cost and schedule for installation.

Next, create the purchase order and get it approved by the level of authority (LOA) process within you organization. After the purchase order is approved and returned to the purchasing agent, ask all the machine tool installation companies to resubmit their bids one more time. This last request is called the *best and final quotation*. This lets the installation companies know that there will be no additional quotes accepted for this installation. After receipt of the best and final quotations, negotiate with the top two companies to get an additional 3 to 10 percent off their best and final quotations. Since you have an approved purchase order in hand, you are in a very favorable position to save the company some money. Offer each installation company something less than what it quoted *and* ask for additional concessions. Even if the installation company doesn't agree to your new terms, you will very likely negotiate a better deal for your company than if you simply accepted a bid at face value. It's always good to show an entrepreneurial spirit on behalf of your employer. It is a rare quality that will not go unnoticed.

5.5   Purchase of the Machine Tool Lubrication Oils and Coolant

The purchase of the machine tool's lubrication oils will be the responsibility of the machine tool buyer. The buyer must choose from the company's purchasing system the right oil suppliers from which to purchase the required machine tool oils. This makes purchasing all the oils and storing them in inventory much simpler. Have at least two or three different suppliers for the machine tool lubrication oils. There will be times during the operation and maintenance of the machine tool when you will need oil and you won't have it. By having more than one company set up in your purchasing system, the needed oils are always just a phone call away. Do this for the coolant of the machine tool, too. You will go through more coolant than oil at any given time.

5.6   Epoxy Painting of the Nonshrink Grout and Concrete Foundation

Epoxy painting of the concrete foundation and nonshrink grout are extremely important. The oils and coolants required for machine tools, today and in the future, will have coolant as the primary substance for extending tool life and the surface finish of the workpiece.

The problem arises when coolants and oils meet the foundation. Over time, these liquids break down the grout that connects the machine tool to the concrete foundation. This is why painting of the foundation is extremely important. If you have a machine tool that leaks oil and coolant onto the foundation and if you haven't sealed the foundation when the machine tool was installed, you can expect alignment problems to appear sometime within the next 10 years. This could require the complete removal and reinstallation of the machine tool with new nonshrink, high-compression-strength grout. This is extremely costly, especially when compared to the cost of applying two-part epoxy paint at the beginning.

5.7 Purchase and Installation of the Machine Tool Metal Chip Bin

The metal chip collection bin, also known as the chip bin or chip hopper, is a metal container that collects the metal shavings and chips created by the machine tool. The chip bin is a piece that the machine tool supplier won't usually provide. It is the responsibility of the buyer of the machine tool to procure the chip bin. Most machine tool suppliers will document on the general arrangement or machine tool drawings that the chip bin is the responsibility of the buyer. See Fig. 5.1 for an example.

If you want to have the machine tool supplier include the chip bin, the supplier will do it, but it will be very expensive compared to buying one outright. The main reason I added this section is that since all machine tool foundations are different, it makes sense that machine tool chip bins will also differ.

See Figs. 5.2 and 5.3. These examples specify, in both the plan and elevation views, what a machine tool supplier usually includes with the machine tool and its normal foundation drawing. As you can see, the machine tool supplier has specified that the chip bin must be approximately 1000 mm × 1000 mm × 1200 mm. This is just an example of what to provide for the machine tool. In these examples, the chip bin is below the factory floor elevation. The chip bin will require either an overhead crane or a forklift to remove it. As you can see again in Fig. 5.1, the chip bin is marked "supplied by buyer," and the machine tool buyer is responsible for figuring out how to

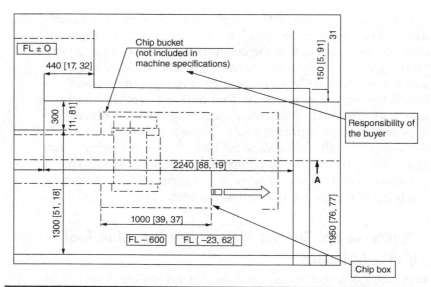

FIGURE 5.1 Chip bin is the responsibility of a buyer example. (*Courtesy of Toshiba Machine Tool, Inc.*)

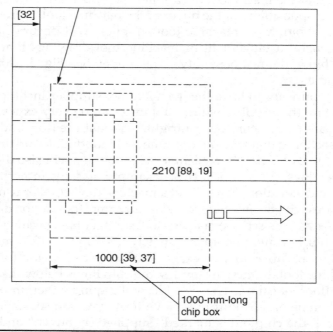

FIGURE 5.2 Chip bin plan view example. (*Courtesy of Toshiba Machine Tool, Inc.*)

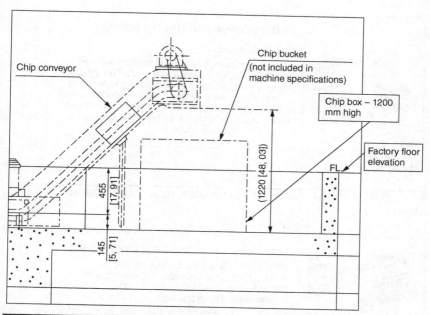

FIGURE 5.3 Chip bin elevation view example. (*Courtesy of Toshiba Machine Tool, Inc.*)

install and remove the chip bin based on the location of the machine tool. There are two options. The first is to make the chip bi-match the machine tool supplier's foundation and general arrangement drawings. The second is to change the supplier's machine tool foundation drawings to accommodate the standard metal chip bin used in your facility. The machine tool supplier won't object to widening or narrowing the foundation to accommodate the buyer's standard metal chip bin.

Shanafelt Manufacturing provides some of the best custom-made metal chip bins in the industry. Shanafelt will custom-make almost any type of chip bin required for an application. The best options Shanafelt Manufacturing offers are the fork rollover metal chip bin or the crane-type skid metal chip bin (see Figs. 5.4 and 5.5). The fork rollover design allows for flipping the chip bin with a rotating device that is attached to the front of a forklift. The crane type, on the other hand, has four lifting points designed into the box for stable control while lifting with an overhead crane. Shanafelt will also manufacture a combination forklift rollover and crane type if needed (see Fig. 5.6). Of course,

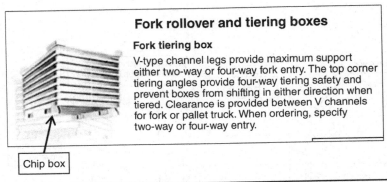

Fork rollover and tiering boxes

Fork tiering box

V-type channel legs provide maximum support either two-way or four-way fork entry. The top corner tiering angles provide four-way tiering safety and prevent boxes from shifting in either direction when tiered. Clearance is provided between V channels for fork or pallet truck. When ordering, specify two-way or four-way entry.

Chip box

FIGURE 5.4   Fork roll over chip box shanafelt. (*Courtesy of Shanafelt Manufacturing, Inc.*)

Crane-type skid platform and box assembly

Designed for loads up to 6000 lb, this box and skid platform assembly has tiering seats, safety tie lugs, and full channel runners as standard accessories. The crane lugs have 2" × 3" holes and are normally made of 3/8" × 4" wide stock, and all top edges are rounded smooth. Lugs are riveted and welded.

Chip box

FIGURE 5.5   Crane type chip box shanafelt. (*Courtesy of Shanafelt Manufacturing, Inc.*)

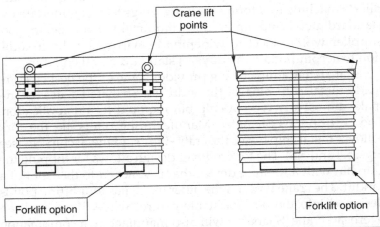

Crane lift points

Forklift option

Forklift option

FIGURE 5.6   Combination crane type and roll over chip box. (*Courtesy of Shanafelt Manufacturing, Inc.*)

the greater the number of options, the more expensive the metal chip bins. The box-type construction allows for the chip bin to be filled more than the standard front, self-dumping, hopper-type metal chip bin. Shanafelt offers many options and will likely have a product that meets your company's needs.

In the previous example in Fig. 5.3, you can actually purchase an additional chip conveyor to place between the machine tool chip conveyor and have the new chip conveyor dump at factory floor elevation. This eliminates using a crane to move to remove the metal bin from the foundation pit.

Some large machine tool suppliers provide the chip conveyor dump in a pit or in the foundation as standard practice. They do this for a couple of reasons. First, it saves floor space by dumping the metal chips in the pit. If the drawings show the chip conveyor dumping at floor level, it will probably increase the overall footprint of the machine tool foundation. Most machine shops don't have the extra floor space to allow the chip conveyor to dump at floor level. If a facility does have the necessary floor space, I would strongly recommend checking out the Shanafelt Manufacturing website at *www.shanafelt.com.*

5.8   Purchase and Installation of the Grout Cans, Grout Forms, and Grout Pour

In the installation specification of the machine tool, the machine tool supplier will detail the recommended nonshrink, high-compressive-strength grout. This is frequently a point of contention among the machine tool buyer, machine tool installation company, and machine tool supplier. There are many different nonshrink grout companies in the marketplace. They are not all the same. The purchase of grout cans and grout forms is usually the responsibility of the general contractor. However, sometimes it will fall to the machine tool supplier, depending on the agreement.

5.9   Purchase and Installation of the Safety Rails of the Machine Tool

There must be a detailed layout drawing, with both plan and elevation views, showing the proper location of the safety rails that encompass the new machine tool. Your EHS manager should review the varieties of safety rails to surround the machine tool and foundation pits before purchasing. If the pit is more than 48 in deep, you must have a toe board at the bottom of the safety rail. Also, if there is an extremely deep foundation where forklift traffic is close-by, be sure to use heavy-duty safety rails designed to stop a 10,000-lb load at 4 mph. There are many different safety rail designs to choose from.

The purchase and the installation of safety rails are usually the responsibility of the buyer of the machine tool. The general contractor who built the foundation or the machine tool installation company can install the rails. The contractor who built the foundation should be required to provide temporary safety rails during the build of the foundation if the contractor won't be providing the final safety rails.

It is important for your team, with the help of the machine tool supplier, to create a layout drawing as early as possible in the process of purchasing the machine tool and building the foundation.

5.10   Overall Installation Schedule

As with the other phases in this book, you should develop an overall schedule for the installation of the machine tool. Figure 5.7 is an example of a standard machine tool installation schedule.

At a minimum, the following tasks should be included in the overall machine tool installation schedule:

1. Confirm the new foundation dimensions.
2. Start installation of the machine tool.
3. Set the electrical cabinet.
4. Set the hydraulic tanks.
5. Set the column.
6. Complete the electrical connection of the machine tool.
7. Make the final mechanical alignments.
8. Test-run the spindle.
9. Train the operation and maintenance departments.
10. Give final acceptance to the machine tool installation.

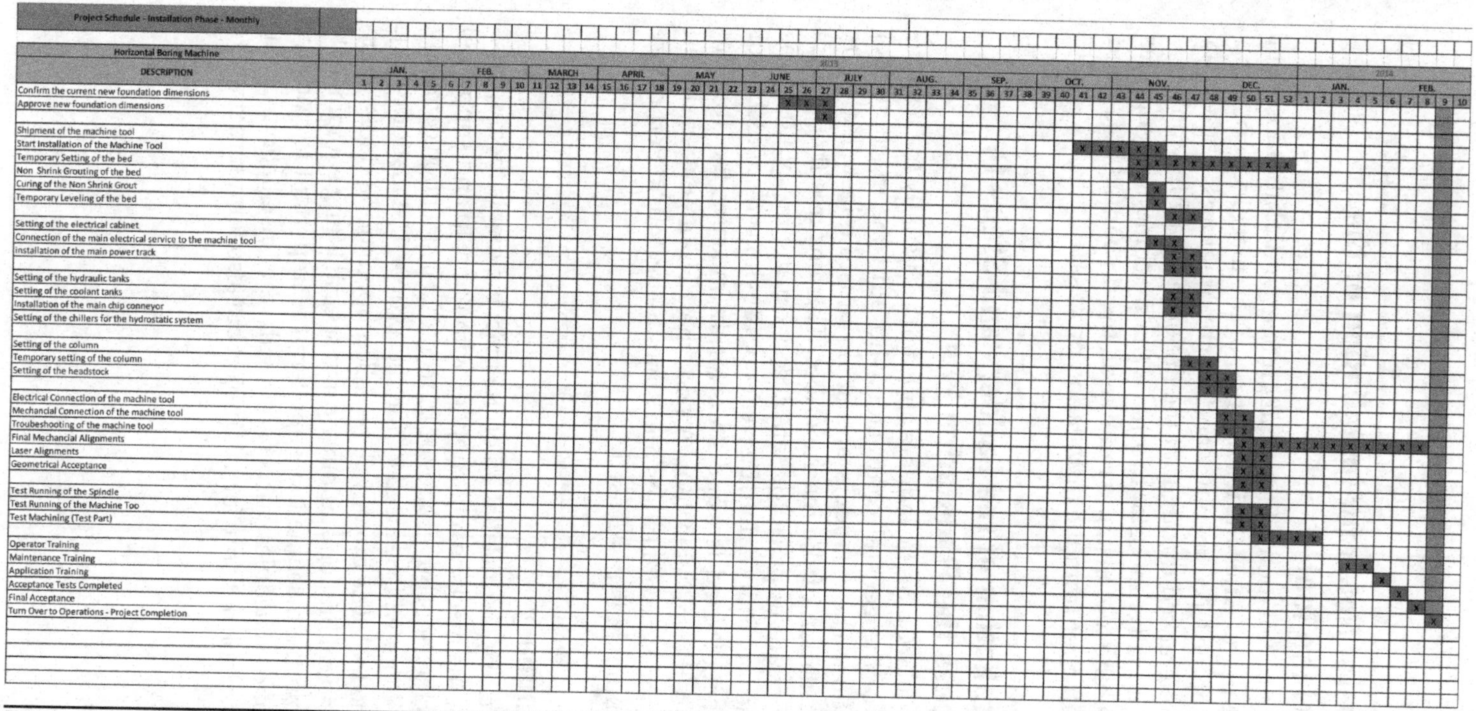

Figure 5.7 Installation project schedule.

5.11 General Notes for Installation of the Machine Tool

After the installation, store all alignment and calibration data in a safe place. After the installation of the machine tool, get a backup of the CNC memory and macro programs. These data will be referenced throughout the life of the machine tool.

It is equally important to have the machine tool supplier and machine tool installer remove all the unwanted boxes and crates left over from the installation process.

5.12 Installation of the Ladders and Stairs to Access the Foundation Pits

The installation of the ladders and stairs is usually the responsibility of the general contractor who built the machine tool foundation. These ladders and stairs must be installed *before* the installation of the machine tool. This ensures the ladders and stairs are in place for safe access during the build of the machine tool foundation. Be clear about who is responsible for purchasing the ladders and stairs for the machine tool foundation and the machine tool. Again, the EHS manager should review the type of ladders and stairs being considered before the equipment is purchased.

5.13 Summary

In summary, the installation of the machine tool is the responsibility of the machine tool supplier with the help and support of the machine tool buyer and/or the contract installation company. Many items will be needed for the proper installation of the machine tool. The buyer will be called upon to ship tools, alignment and calibration equipment, and special lifting equipment to and from the machine tool supplier. It is important to track the costs and schedules associated with the installation of the machine tool, not only for this project but also for future projects.

CHAPTER 6

Preparation Phase

6.1 Preparation Phase

By now, it should be clear that we are using the terms *machine tool manufacturer, machine tool supplier,* and *original equipment manufacturer* (OEM) interchangeably

In the preparation phase, we will highlight potential problems that will only serve to delay the design, purchase, and installation of the machine tool. We will also make some recommendations on how to prevent unthought-of and unknown problems during the purchase and installation of the machine tool. You will find examples of challenges that have caused me great frustration in my career. Carefully consider each and the preventive steps as key elements of your preparation.

6.2 Chip Conveyor Requirements When the Chip Conveyor Is Supplied by the OEM

The chip conveyor can be supplied by the OEM, but it can also be purchased from a variety of other vendors. The chip conveyor usually isn't something the OEM manufactures, but if you get it through the OEM, you will really be buying the controls, wiring, PLC logic, and machine tool layout, along with the new chip conveyor. You will receive a complete turnkey chip conveyor solution for your new machine tool.

The tradeoff for buying the chip conveyor from the OEM is that it comes at a substantially higher cost than outsourcing the purchase to an independent vendor. For the same money, you can purchase a much heavier-duty chip conveyor with the same options as the one from the OEM. In my experience, if you purchase the chip conveyor from the OEMs, it will cost 10 to 30 percent more than purchasing from a chip conveyor manufacturer, and it will not include all electrical and mechanical connections or the layout of the new chip conveyor.

6.3 Chip Conveyor Requirements When the Chip Conveyor Is Not Supplied by the OEM

If you purchase the machine tool without a chip conveyor, here are some things to consider: First, make sure the chip conveyor installation schedule is on track to go in before the installation of the machine tool. It may be impossible to install the chip conveyor after the machine tool bed has been installed. The anchoring and leveling of the chip conveyor will be extremely difficult without clear, unimpeded access to the side and underneath of the base of the chip conveyor.

Second, consult with the OEM about the particulars that will allow for the needed space for the main control cabinet with these controls. Whether the chip conveyor supplier, OEM, maintenance electrician, or facilities electrician will install these components is a matter that must be settled well in advance, to avoid delays.

Third, make sure the chip conveyor supplier includes the proper fuse block with fuses or a circuit breaker. You will also need proper-size reserving motor starter for this chip conveyor, which should be included in the electrical cabinet. Confirm that the power source feeding the chip conveyor is supplied from the same electrical cabinet as the machine tool. This is done so that when the machine tool is powered off and locked out for maintenance or facilities issues, the power to the chip conveyor is also off. If the power to the chip conveyor comes from a source *other* than the main control panel, you have a serious safety issue.

Fourth, only purchase a chip conveyor equipped with reverse-direction capability on the belt of the conveyor. This is essential because the operator will forget to turn on the chip conveyor for an extended period, which causes metal chips to accumulate on the conveyor belt. This accumulation can cause the chip conveyor to hang up or trip the main breaker when reactivated.

It is also essential that the head shaft end of the chip conveyor have enough clearance for a large ball or mass of metal chips to pass through the small opening at the end of the chip conveyor.

Today's advanced CNCs and programmable logic controller (PLC) controls operate most machine tools on the market. The machine tool is able to turn the chip conveyor on and off by an M code in the part program. If you purchase the chip conveyor from the OEM, you're likely to receive with it the controls for turning on and off the chip conveyor in the part program. This PLC code is usually added during the build of the machine tool at the factory. If you don't purchase the chip conveyor from the OEM, you won't have the option of turning the chip conveyor on and off unless you specify this option in the beginning of the purchase of the machine tool.

Another issue that you may run into is the need to tie the machine tool chip conveyor into another chip conveyor system. Some very large machine shops have many large, underground metal chip conveyor systems that collect all the metal chips produced in a facility and move them to a centralized point. When the main system shuts down for a maintenance or facilities issue, it actually turns off the chip conveyor at the machine tool. Turning the conveyor off at the machine tool prevents all the chips on the chip conveyors from piling up on the main chip conveyor system while the main system is down. This synergy can occur only if you and your maintenance and facilities group interlock the chip conveyor at the machine tool with the main chip conveyor system. This is done by adding a proximity switch or a zero-speed switch onto the main chip conveyor to check for movement of the main system belt. Whenever this belt is not moving, it sends a signal to the machine tool CNC and PLC to inform it that the main chip conveyor system is down. It will become apparent to every department quickly if you don't interlock the two systems. The first time the main system shuts off and someone has to remove deck plates and covers to clean out the metal chips, you will hear about it, and not in a good way.

Another extremely important interlock is the interlocking of the spindle to the chip conveyor. If the operator forgets to turn off the chip conveyor for 10 min after the spindle of the machine tool is stopped, the chip conveyor will then automatically turn off.

It is important to allow the operator to control the chip conveyor from the main panel. The best way to do this is to install four push-buttons to the main control panel or main operator pendant for turning on and off the chip conveyor and forward and reserve of the belt. You don't want the operator climbing up and down the operator platform each time he or she needs to start the chip conveyor.

6.4 Chip Deflectors for the Chip Conveyor and Other Sections of the Machine Tool

Chip deflectors or "baffles" are pieces of steel or rubber that deflect metal chips away from the operator and redirect them to the main chip conveyor of the machine tool. Chip deflectors fill the unused space between the machine tool and chip conveyor. The OEM is able to provide chip deflectors for your machine tool if you request it.

The more the work is performed by the machine tool, the more metal chips are made. You will need heavier-duty chip deflectors that can withstand the constant abuse caused by these powerful metalworking machine tools.

Once you have purchased the chip deflectors from the OEM, have your local fabrication company measure and install them after the machine tool has been installed and the chip conveyor is fully operational. Have the installers return three or four times to readjust them. This is done because the concrete foundation and chip conveyor will not be perfectly installed to within normal machining tolerances and alignment standards. For large machine tools that require coolant, the chip conveyor is normally pitched or sloped toward the main coolant tank for efficient drainage. The combination of the foundation and the slope of the chip conveyor may exceed one-half in elevation over the entire length of the chip conveyor. Fortunately, the through-holes for the deflectors are slotted for adjustment if installed after the machine tool has been installed.

6.5 Main Power Source for the Machine Tool

The main power source to the machine tool will be the responsibility of the buyer of the machine tool. The main power source to the machine tool should be purchased well in advance of the installation of the machine tool. It could take months to manufacture a large-ampacity circuit breaker, bus duct switch, or large safety switch disconnect. If you have a bus duct system that powers the machine at your facility, be sure to purchase a circuit breaker–type bus duct switch for this bus duct. The reason for the circuit breaker bus duct switch is that the bus duct is usually at least a 20 ft in elevation. It is extremely difficult and time-consuming for the operational department to change fuses at 20 ft in the air.

6.6   Secondary Power Source for the Machine Tool

The secondary power source for the machine tool is also the responsibility of the machine tool buyer. The secondary power is otherwise known as the *single-phase power*. It is normally a 115-VAC single-phase power source. It is advisable to install at least four or five 115-VAC outlets in the machine tool pit foundation to run sump pumps, hand tools, and additional lighting.

6.7   Compressed Air

Most machine tools require compressed air for proper operation. The machine tool supplier will detail exactly where they you should supply compressed air to the machine tool on the machine tool foundation or general arrangement drawings. See Fig. 6.1 for an example of where the machine tool supplier wants compressed air located.

Some of the newer machine tools will have compressed air that operates through the headstock and by an M code in the part program. The compressed air is directed at the tool as it performs the work, which keeps the cutting tool cool, directs the metal chips away from the tool, and blows them into the chip conveyor.

Always install an in-line "T" fitting in the main air connection for additional connection in the future. It is always easier to add flexibility into any system at the beginning of an installation than to change it later when the machine tool cannot be shut down due to machining requirements and schedules.

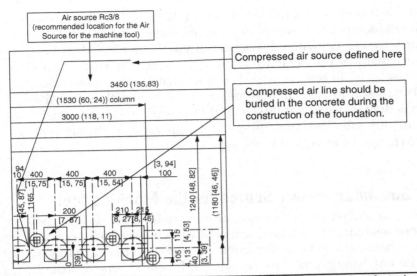

Figure 6.1 Compressed air line detail. (*Courtesy of Shanafelt Manufacturing, Inc.*)

6.8 Water

If the machine tool has a coolant system and a large foundation, it usually has a large coolant reservoir that holds more than 500 gal of water and coolant. During the operation of the machine tool, not all the coolant will make its way back to the coolant tank. The coolant that doesn't make it back to the coolant reservoir will evaporate quickly, so it is important to have access to a water source within a reasonable distance, to supply water to the coolant tank as needed.

6.9 Chilled Water

If your machine tool is designed with a tube-style water-cooled chiller, it will need chilled water. This water is supplied by the facilities department. Some facilities won't have a chilled-water source. The OEM should detail what type of chiller is installed on the machine tool for proper cooling of the hydrostatic and headstock of the machine tool. Ask this question during the procurement phase, but also revisit the issue during the preparation phase since the OEM could change the design without informing you.

6.10   Coolant/Oil Skimmers

These skimmers are usually not included in the purchase of the machine tool. The oil skimmer separates the oil from the coolant by means of a disk or belt.

Disk-type skimmers are oil skimmers that rotate a disk-shaped medium through the coolant. Oil is then wiped off and discharged into a collection container through plastic or steel piping. See Fig. 6.2 for an example of a disk skimmer that you can purchase for the machine tool. The disk-type skimmer can also be discharged into the machine tool's sump pit or tanks and manually collected by the maintenance mechanic, technician, or machine operator. The collection container or sump pit could also be tied directly to the coalescer for additional filtration. It is important to consider *reach*, which is the portion of the disk that actually gets immersed, when selecting a disk oil skimmer. The least amount of surface area of the disk in the fluid means less oil removed. Disk-type skimmers are often inadequate to handle fluctuating fluid levels within the coolant tank.

Belt-type oil skimmers use an endless belt of corrosion-resistant steel or synthetic medium, which is lowered into the

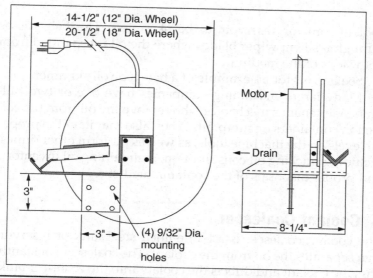

FIGURE 6.2   Disk type oil skimmer. (*Courtesy of Encyclon, Inc.*)

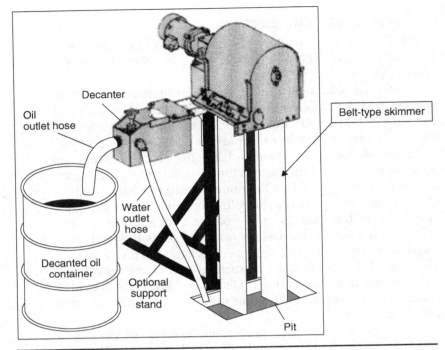

Oil outlet hose

Decanter

Belt-type skimmer

Water outlet hose

Decanted oil container

Optional support stand

Pit

FIGURE 6.3   Belt-type oil skimmer.

coolant tank or reservoir to be skimmed. The belt passes through resilient wiper blades where the oil is separated from both sides of the medium.

See Fig. 6.3 for an example of a belt-type oil skimmer.

Many larger machining companies have one or two full-time maintenance mechanics who remove the oil from the collection containers or sump pits. They also maintain the proper oil levels on the machine tools, as well as perform other duties, including tracking oil consumption, maintaining oil inventory, and proper disposals of the contaminated oils.

6.11   Coolant Coalescer

The coolant coalescer is a baffle-lined steel tank or reservoir that separates the oil from the coolant. The coalescer constantly stirs the coolant and directs the coolant into the coalescer tank. Coolant passes through the baffles and separates the oil from

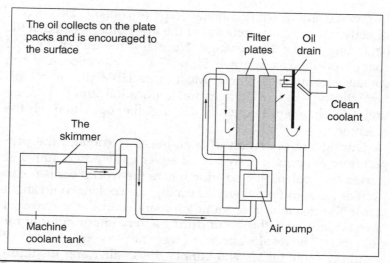

The oil collects on the plate packs and is encouraged to the surface

Filter plates

Oil drain

Clean coolant

The skimmer

Machine coolant tank

Air pump

FIGURE 6.4   Tramp coalescer-type oil skimmer. (*Courtesy of LNS-Turbo, Inc.*)

the coolant. The oil is then channeled into a collection container or to the sump pit. The main difference between the coalescer and oil skimmers is that the coalescer is a pneumatic device and the oil skimmers are mechanical, driven by an electric motor. The coalescer is nearly maintenance-free, whereas oil skimmers will eventually need electric motor repair or replacement, as well as changes of belts and brushes. See Fig. 6.4 for an example of a coalescer oil skimmer.

Keep in mind that the coalescer oil skimmer will consume additional floor space around the machine tool foundation and will need compressed air to operate.

6.12   Coolant Wall and Enclosure

In the purchase of a medium or large HBM equipped with a tool changer, coolant, or high-pressure coolant, the coolant wall and enclosure should be considered from the beginning. The containment of the coolant within the workspace of the machine is important for proper long-term operation and safety. Another reason to consider installing a coolant wall is that during the operation of a tool changer, after the tool holder is removed from the spindle, air is forced through the spindle to remove any unwanted metal chips or coolant. This occurs directly following the removal of the tool holder located at the beginning of the tool change. Normally the automatic tool change position is between 5 and 7 ft in elevation and can spread coolant everywhere. Most large HBM machine tools have this type of tool changer and application, and this operation will happen frequently during the operation of the machine tool.

During the operation of high-pressure coolant, the programmer or operator may have inadvertently programmed or started the coolant in a position where the workpiece is not in front of the spindle. When this occurs, the coolant could shoot across the machine tool foundation and into the walkway. I have actually seen this occur during a very important tour for customers. The results afterward were not so good.

Today some OEMs will supply the smaller and medium-sized HBMs with total enclosures around the machine. Be sure to ask your OEM if it will be able to supply these enclosures for your machine tool during the procurement phase or after installation of the machine tool if your company didn't have enough capital funding at the beginning of the project.

6.13 Touching Up the Epoxy Coating of the Foundation

After the installation of the machine tool on its foundation, it may be necessary to touch up the epoxy paint on the machine tool foundation. During installation, the OEM or the machine tool installation company will probably scar and damage the foundation. Remember to epoxy-paint the deck plate and any structural steel cover that was added after the installation.

6.14 Lighting in the Foundation Pits

If your foundation has a pit or area that is covered with bar grating or deck cover, which you have to access for maintenance or facility reasons, install explosive-proof lighting in the pit. It is also advisable to supply power to these lights from the plant operations services, not from the main machine control panel. If machine tool has to be powered down, access to the pit is still essential. By having the power source at the building's operational service, it allows you to access the pit while the machine tool has been electrically locked out. If your facility has an emergency backup power service, and assuming someone is keeping it ready for service, have at least one or two of the lights parallel to the emergency power source. If the entire facility should lose power, this will allow work in the pit to continue or people to exit the pit safely.

6.15   Isolation Transformer

During the purchase of the machine tool, the OEM should let you know if the machine tool will require an isolation transformer. In some cases, your company's facilities or maintenance department will determine whether an isolation transformer is needed also. The job of the isolation transformer is to even out a sporadic and inconsistent voltage flow to the machine tool. It is a necessity if your company or utility supplier doesn't have a constant voltage. Many factors can cause this fluctuation, and they are difficult to diagnose and usually impossible to fix. The machine tool will not tolerate voltage fluctuation for very long before it sustains serious damage. To be safe, consider the purchase of an isolation transformer a must. The isolation transformer is there to protect the machine tool from voltage spikes and to buffer the noise of other machine tools around your machine tool. This device also protects the electrical service flowing away from the machine tool. Most machine tools have variable-frequency drives (VFDs) for axis movement. Sometimes these VFDs put unwanted noise back into the electric power distribution of the facility. The isolation transformation helps reduce the amount of noise produced by the machine tool going back into the electrical distribution.

6.16   DNC Software

After the machine tool has been installed and the direct numerical control (DNC) connection completed, you should see if it actually works. Also, there should be a written procedure for uploading and downloading the part program to the CNC. This procedure should be documented, saved on the company's network in a safe location, and located in the work instructions

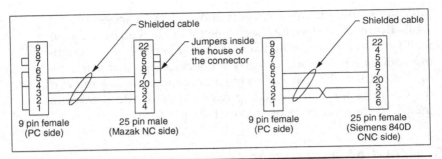

FIGURE 6.5   RS232C connection examples.

for the machine tool. It should also be posted at the machine tool for all to see. DNC is a software program and system that allows part programs to be downloaded to the CNC of the machine tool. Most DNC software programs can be downloaded via the Ethernet port in a cable or wireless connection. In the past, most DNC connections used the RS232C port of the CNC. This is the 9-pin serial port on the back of a standard desktop personal computer. Figure 6.5 shows an example of two RS232C connections for a Mazak CNC and Siemens CNC. These two examples detail the connections between the 9-pin serial port located at the personal computer and the 9- or 20-pin connections at the Mazak CNC or Siemens CNC, which is located at the machine tool main control panel or at the main operator's panel.

Per the example, make sure that the cable between the personal computer and the CNC is a shielded cable. Also Fig. 6.5 details some examples of needed hard-wired jumpers that may be required for proper connection between the personal computer and CNC.

Some manufacturing companies in the industry prohibit the installation of CNC on the internal company network. Some of these companies have a standard operating system that all their computers use. Most newer CNCs today are Windows-based, and your IT department may not allow networking the new machine tool to the main network. Contact your IT department before starting the purchase and installation of your DNC software to the machine tool, to determine whether the CNC for the new machine tool can be connected to your company's network.

6.17   Direct Numerical Control Procedure

After the machine tool has been installed and the DNC connection completed, you should see if it actually works. Also, there should be a written procedure for uploading and downloading the part program to the CNC. This procedure should be documented, saved on the company's network in a safe location, and located in the work instructions for the machine tool. It should also be posted at the machine tool for all to see. The physical wiring connection of the DNC should also be saved and stored in a safe place for ongoing maintenance.

When you are sending and receiving a part program to the Fanuc CNC, there are two specific processes. The first is called *punching* (sending the part program out of the CNC) and the second, *reading* (receiving the part program to the CNC). When you are using a Siemens CNC, the terminology is simply *data in* (receiving the part program to the CNC) and *data out* (sending the part program out of the CNC). All parameters for the proper sending and receiving of part programs should also be documented during the preparation phase.

6.18   PC at the Machine Tool for Network Connection

Depending on your company's IT requirements, you may be required to install a PC at the machine tool for network connection. Most companies don't want the CNC on the company network. The IT department is concerned that someone could access the company's secure files through the CNC. IT will require you to install a personal computer close to the machine tool for part program file transfer through the RS232C protocol. This transfer of the part program and back data to the CNC is usually made through the PC's serial port nine-pin connection.

6.19   Main 480-VAC Breaker

The purchase, installation, and connection of the main three-phase 480-VAC circuit breaker, which isolates the 480-VAC operating power to the machine tool, are the responsibility of the machine tool buyer. Keep in mind that in the United States, the power requirement to run certain machine tools can also be

220 VAC. Because of the lead time needed to purchase, manufacture, and install the main breaker at your facility, the likelihood will be small of getting the main breaker at the same time as there is a scheduled power outage to the production facility where the machine tool is located. Most likely you will have to schedule downtime in other areas of the factory to install this breaker because the 480-VAC power servicing the machine tool will also be powering other machine tools and equipment in the facility. To add this main breaker will cause you to most likely shut down the entire power panel or bus duct.

6.20 Deck Plate, Checkered Plate, and Bar Grating around the Machine Tool Covering the Machine Tool Foundation Pits

The deck plate, checkered plate, and bar grating are the structure steel that surrounds the machine tool. These should be installed as early as possible after the foundation is completed, for two reasons. The first is for safety. Open pits and trenches in the foundation while the machine tool is being installed are a clear safety hazard. The second is that you should know as early as possible of any obstruction between the machine tool and the deck plate, checkered plate, and bar grating. Sometimes this combination of deck plate, checkered plate, and bar grating is called *structure steel* and is associated with the machine tool.

6.21 Laser Alignment Checks

The initial laser alignment of new machine is performed using the buyer's laser alignment equipment, if available. This same equipment will be used to check the laser alignments for years to come. The laser acceptance checks should be performed at the location where the tool is cutting the metal. This means that the laser setup should be performed at the tool post on a lathe and out of the spindle when on an HBM.

It is also worth noting that if you have fully hydrostatic machine tools, cycle the machine tool for approximately 1 h to evenly distribute the heat through the bed sections or column. This means that you should constantly move the axis, which is to be laser-checked 1 h before you perform the laser accuracy check. You can perform this function by writing a short program for the CNC to maintain continuous movement, or you can move the machine tool manually by jog mode using the main control panel of the CNC.

Also, turn off the heating, venting, and air conditioning (HVAC) systems and any other air movement ventilation systems close to the machine tool while making these laser checks. If possible, lock out and tag out the electrical source to any large overhead roll doors within 40 ft of the machine tool. The change in temperature each time the roll door opens will definitely affect your laser alignment checks. It would also be difficult to perform laser alignment checks in a facility where the temperature changes rapidly within the period the alignment checks are performed.

The larger the machine tool, the longer these checks will take. You should account for this time in your overall machine tool installation schedule.

6.22   Final Machine Tool Acceptance

During the final machine tool acceptance detail, list any issues that remain from the machine tool installation. This list is called the *final acceptance punch list*. Due to machining requirements, it may be necessary to accept the machine tool and start production. The more complex the machine tool, the larger the final acceptance punch list. Since you have the OEM on the hook for issues related to the acceptance punch list, it is easier to add more issues. The final machine tool acceptance should be signed off by the project engineer or by the department manager, who will ultimately be responsible for the machine tool.

6.23   Machining Test

A machining test begins with a test workpiece similar to the type you plan to machine during the normal operating production. Use the same tools and tooling that you plan to use during normal production. It is of the upmost importance to require the OEM to perform some sort of machining test during the runoff and acceptance of the machine tool. The machining test should include a horsepower check, heavy cuts, as well as rough and finish machining operations. Also, duplicate the very machining applications and profiles for which you plan to use the new machine tool, with the OEM present.

During the machining tests, it is important to check the run-out radially and axially of the machine processes using a 1000- or 1-μm indicator. This forces the machine tool supplier to guarantee that it is producing a solid and accurate machine tool.

6.24   Develop and Test a Part Program

Nearly every machine tool has a different CNC. It is important to create a part program as early as possible to check the operation of the machine tool. The test part program should contain all G and M codes associated with the machine tool. Each code should be checked for functionality and operation. This type of testing can be performed during the operator and application training of the machine tool at the buyer's facility.

6.25   Teleservice Provided by the Machine Tool OEM and Installed and Tested by the Machine Tool Buyer

Teleservice is a direct Ethernet or telephone connection that allows the OEM to access the machine tool CNC from the OEM's location. This option enables the OEM to troubleshoot the machine from a remote location. It also allows the OEM to modify and install the machine tool CNC and PLC, if needed. Teleservice is something that is installed and checked at the end of the installation, though sometimes it is overlooked.

6.26   Warranty

The warranty period of the machine tool begins with the sign-off of the final acceptance of the machine tool. One of the most important things to know about the warranty period is the final date. If the machine tool breaks down during the warranty period, it is important to understand that the OEM's responsibility is usually limited to replacement parts. The service required to remove and change the damaged parts and the cost of getting the service technician to the location will not be covered. This is something that should be addressed in the purchase of the machine tool, but it is sometimes overlooked.

6.27 Conduit (EMT, IMC, Rigid, and PVC) Required for the Machine Tool

The installation and the connection of the DNC system are another expense that falls to the buyer. The buyer is responsible for the physical connection of the wireless, Ethernet, or RS232C connection of the company's computer network to the CNC of the machine tool. The proper type of conduit and cable to and from the machine tool must be installed. More importantly, this is the same for the main 480-VAC three-phase service and 110-VAC single-phase service to the machine tool.

When you are installing or running conduit to the machine tool or a heavy piece of equipment, it is important to understand the differences between the four most commonly used types of metallic and nonmetallic conduit in an industrial setting for a machine shop or heavy manufacturing facility. These differences have been spelled out before and are based on my recommendations.

Issues	Recommendations
Rigid metal conduit (RMC). This conduit is the heaviest and thickest wall option of all the conduit options.	I would recommend using rigid metal conduit where there is a risk that the conduit could be damaged by a forklift, crane, pallet jack, metal pallet, or rigging equipment. If you need to install conduit in a place where there is a possibility that someone might step on the conduit or use the conduit to support additional piping or conduit, I recommend you use RMC. It is the most expensive option because of its heavy-duty steel wall and because it is more difficult to install.
Intermediate metal conduit (IMC). This conduit is a thick-walled galvanized option that is not as heavy-duty as RMC. IMC is the middle metallic option between RMC and EMT.	I would recommend using intermediate metal conduit on the walls of the machine tool foundation and inside the trenches of the machine tool. IMC is less expensive than RMC and more expensive than EMT. Also IMC is easier to install than RMC and just slightly more difficult to install than EMT.

Issues	Recommendations
Electrical metal tubing (EMT). This type of conduit is the most lightweight and thinnest wall-type metallic conduit. Also EMT is the cheapest and easiest type of conduit to purchase and install.	In a machine shop or heavy-duty facility, I would recommend using EMT only in places where there is no chance of damaging the conduit with a crane, forklift, pallet jack, wood or metal pallet, maintenance cart, metal cart, or any other moving vehicle. Do not install EMT near a roll-up or overhead door. EMT is for light manufacturing facilities and commercial applications only.
PVC—nonmetallic rigid conduit. PVC conduit is a type of thick-wall plastic conduit which is used mainly in underground-type applications.	In a machine shop, coolant and water connections to the machine tool can be made using PVC pipe. Also PVC pipe can be used between the chip conveyor and the drainage system to the main coolant tank. Use PVC piping in applications where rust is a concern.

One other warning is needed in regard to the types of conduit mentioned above: never install conduit on the top of the concrete foundation and in a location where there will be foot traffic. Conduit of any type should be installed in this way only if there is no other option to install the conduit safely. Always be open to chipping out the concrete to approximately 3- or 4-in depth, installing the conduit, and repouring the concrete or grout. Even though this will be a more expensive option, it will not be a safety issue with the conduit lying on the floor and posing a trip hazard.

The purchase and installation of the metallic and nonmetallic conduit for the machine tool foundation and around the machine tool most likely will be completed during the installation of the machine tool. The conduit between the main circuit breaker and main control panel to the machine tool should be rigid metallic conduit. Even though there is a low risk of damage, you don't want the main power distribution source to the machine tool damaged.

6.28   Power Track or Wireway

The OEM might require the buyer to install a power track or wireway for wiring between the main control panel and the machine tool. The power track or wireway goes into the machine tool foundation and under the trench cover around the machine tool. If your machine tool has coolant capability, the OEM will ask that the wiring be covered as much as possible. Be sure to double-check the machine tool specification and general arrangement drawings for any mention of additional power tracks or wireways that are required.

6.29   Head Attachments (Head House or No Head House)

During the installation of the machine tool, be mindful that the head attachments should be able to be installed and removed within the actual travels of the machine tool. The loading and unloading of the head attachment can be performed with cycles in the CNC and in automatic or MDI mode. Avoid installing and removing the head attachments with an overhead crane each time and in manual operation mode. Don't let the OEM persuade you to disallow the head attachment within the actual travel of the machine tool. The loading and unloading of every head attachment should be performed during the preparation phase. This function should be performed at least 5 to 10 times per head attachment. Cycling of the head attachments will be a good training exercise for the OEM, maintenance and facilities, and operations. It is also important to detail the steps needed to recover normal operation if there is an emergency stop condition, machine tool alarm, or CNC alarm during the installation or removal of the head attachment.

6.30   Sag Compensation

Most of the new CNCs have the capability of sag compensation, a function that electronically compensates for the mechanical deflection associated with some machine tools. This option is usually purchased from the CNC manufacturer separately and not from the OEM. Sag compensation is needed when extending the ram of an HBM. The ram of an HBM is quite a large, heavy mechanical device that droops as it extends outward. Sag compensation makes up for the mechanical deflection of the drooping by slightly moving the vertical axis of the HBM up or down.

6.31 Temperature Compensation

Temperature compensation is needed for the machine tools installed in locations where the temperature changes rapidly and drastically. The temperature compensation parameters may need to be adjusted for summer and winter if the facility doesn't have an HVAC system. Sometimes temperature compensation is needed for hydrostatic machine tools. These have a large temperature gradient between the bed and the column of the machine tool. When the column moves down the bed, the temperature of the bed is much lower. This can cause machining errors on the workpieces. Most likely, the temperature compensation parameters will not be functional when you are purchasing the machine tool. It is a separate add-on option that comes from the CNC manufacturer.

6.32 Checking of the Hard and Soft Overtravel Switches of the Machine Tool

The proper operation of the hard overtravel switches for each axis of the machine tool should be confirmed after the machine tool has been installed. A hard overtravel switch is an actual mechanical or electrical switch. There are two reasons to check the hard overtravel limits. First, the overtravel switch stops the movement of each axis before it runs out of mechanical and electrical travel in each direction. The machine tool has a positive and negative direction for each axis.

Second, it verifies the maximum travel of each axis. The OEM has sold your company a machine tool that claims a specific amount of axis travel. This exercise confirms that the amount of axis travel matches the machine tool specification.

The soft overtravel switch for each axis is only activated after the machine tool has been referenced for that axis. Unlike the hard overtravel switch, the soft overtravel switch is a software switch in the CNC parameters. There is a positive and negative software limit for each positioning axis. As with the hard overtravel switch, the soft overtravel switch should be checked to confirm that it is working properly.

The amount of travel of each axis for both the hard and soft overtravel switches should be documented and kept in a secure place for future reference. If the distance between each overtravel limit switch is shorter than what was claimed, ask the OEM to move the overtravel limits per the machine tool specification. If the OEM can't do this, request a discount on the final payment.

6.33   Basic Reference of the Machine Tool

The reference axis of each machine tool should be documented and shown to the buyer of the machine tool after installation. Sometimes the order in which each axis is referenced matters and needs to be detailed and given to the each operator, machinist, and maintenance technician. The machine tool has to be referenced before the machine tool can be run in automatic or manual data input (MDI) mode. In referencing the machine tool prior to movement of the axes, it is important to know that the software limits of the machine tool are not functional.

6.34   Start-Up of the Machine Tool from an Off State

The start-up of the machine tool from the off position must be documented and shown to the buyer of the machine tool after the installation. First the main disconnect is switched on. Then the main disconnect on the machine control is engaged. Next the CNC is activated. After the CNC boots up and all the control alarms are reset, the hydraulics of the machine are actuated. Now the machine tool is ready to be referenced and ready to run.

Document the three-phase power by checking the voltage phase to phase across all three legs, and from phase to ground between all three legs. The voltage will vary from the morning to the afternoon. It will also vary with the addition of equipment at the facility and with the distance from the power source. The farther away the power source, the lower, most likely will be the voltage.

6.35   Lock Out and Tag Out Procedures and Disconnect for the Machine Tool

Before locking out and tagging out the machine tool at the main disconnect, some things should be reviewed before you turn *off* the machine.

First, turn off the hydraulics of the machine. Second, depress the emergency stop switch. The machine is now ready to be turned off via the main disconnect on the control panel of the machine tool. Third, either lock out the control panel disconnect or turn off the main disconnect, which feeds the main disconnect on the machine control panel. The final step is to lock out and tag out the main disconnect to safely work in the machine tool's control panel.

It is important to post a sign on the main control panel next to the main disconnect that details from where the main power source is being fed. Many times in my career we have turned off the power to the main control panel without knowing the precise location of the power source.

6.36   Backup of All the CNC Data

After the final sign-off of the acceptance of the machine tool, back up of all the CNC data. This function should be performed with the OEM and the installation team present. The team that I am referring to comprises the electrician, mechanic, technician, and project engineer. This function is extremely important. Your company must have the latest data backup in case of a failure of the CNC main processor or the PLC's main memory. This is where it is acquired. These data should also be saved on the CNC, if possible. Sometimes the reloading of the CNC data will fix unexplainable problems with the machine tool.

6.37   Viewing Of and Access To the PLC Ladder Logic

During the installation of the machine tool, ask the OEM's electrician to detail the steps needed to access the ladder logic by the main control panel or with an external laptop computer. It is important for you to document this information because it has to be relayed to every shift that maintains the machine tool.

The viewing of the CNC/PLC ladder logic is ideal for the maintenance technician in troubleshooting the machine tool. Access to and viewing of the ladder logic will require additional software that will not come with the purchase of the machine tool.

6.38 Summary

During the preparation phase, we have detailed the necessary steps and actions that come after the installation of the machine tool. By following these instructions, which are influenced by my own past mistakes, there should be minimal delay in the start-up of the machine tool. Conversely, not following these instructions closely and not providing detailed documentation to the proper people for future use will most assuredly lead to extremely problematic situations.

CHAPTER 7

Start-up Phase

7.1　Start-up Phase

During the start-up phase, there will be some debugging of the machine tool and all its functions. Every large machine tool installation presents many start-up issues. From burning out new servomotors to damaging the linear scales, to replacing the CNC control, to replacement of a coolant pump, problems and downtime are to be expected.

Remediation of all start-up problems is the responsibility of the OEM. From the buyer's perspective, it is crucial that these issues be resolved as quickly as possible. However, in most situations, the process is anything but quick. On the contrary, it is time-consuming and costly to the buyer because the new tool isn't producing. It is also costly to the buyer because the employees of the company who are responsible for the project must oversee the entire remediation process, from the identification of the issues to the final resolution. It is common for the problems of the start-up phase to consume 50 to 75 percent of the workday of each employee involved. The reason for this continuous oversight is to ensure that the OEM solves the start- up problems as quickly as possible. The operational department is waiting to take custody of the tool to engage it in the manufacturing process.

7.2　Machining Test and Test Parts

Most OEMs are used to performing a machining test or some other type of test as part of the acceptance of the tool. Usually, the OEM has a standard test part selected that is always used for this purpose. This test should never be waived as it is an essential part of the start-up phase. The buyer of the machine

153

tool should also perform the same machining processes, using their normal day-to-day machining process to create the test part. The machining test provides a clear understanding of how to run the machine tool, as well as what new tooling or part program changes will be needed.

7.3   Acceptance Tests

Every function of the tool must be checked and signed off during this phase. It is the responsibility of the entire team to test every function of the main operator control panel and main control panel during the acceptance test. This also includes any switches, push-buttons, or lights associated with the tool.

Every emergency stop should be individually tested by pressing each emergency stop button once to confirm that it does disable the outputs of the PLC and CNC, as designed. Next, disengage the emergency stop button and reset all alarms on the CNC. Then press the next emergency stop button, and repeat these steps until all emergency stop buttons have been thoroughly tested.

7.4   Final Machining Test

The final machining test differs from the standard machining test in that the part to be machined is one that your company actually produces for sale in the normal course of its operation—not one of the OEM's choosing. The OEM should attend during both tests to witness the tool's performance under both circumstances.

The OEM should then remain on-site after the final acceptance of the tool for at least a month to be readily available to correct any other issues that may arise with the tool. This arrangement should be included in the purchase contract at the beginning of the purchase, leaving some room for negotiation in the event the installation falls behind schedule.

7.5   Continuous Running Test

After the tool is installed at the buyer's facility, the continuous running test is performed. As the name suggests, the tool is made to run continuously for a prolonged, predetermined period—usually between 6 and 24 h. This test is designed to push the tool's capability to its limits. It involves turning on and off all spindles, running the spindle up to maximum speed, rapid and programmed movement of all axes, automatic installation and removal of all head attachments, automatic changing of every tool in the tool changer, and turning on and off of the coolant. The continuous running test is optional. It may or may not be included in the original purchase contract.

Always rope off the area to prevent accidental entry to the area where the tool will travel. If the tool alarms out or hangs up during the test, the OEM must detail, in writing, the cause of the incident and then create a corrective action report outlining the actions necessary to remedy the problem. After the corrective action report is approved by the buyer, the OEM will repeat the test. This process continues until the machine tool passes the entire continuous running test.

7.6   Uptime Accountability of the Machine Tool

As with downtime, the uptime of the tool must be documented. This documentation begins immediately when the tool goes into service. If the tool breaks down continually during the warranty period, if it develops a major machining issue due to the machining test, or if the machining of the first real workpiece is substandard, request that the OEM provide additional spare parts and technical support.

7.7   Machine Tool Supplier Support during the First Year

Have the OEM supply one technician for the first year to help maintain the tool. It is not practical for the buyer to hire a dedicated technician for this purpose. If the purchase involves more than one machine tool at a time or if the installation of each machine tool is spread over a span of 3 or 4 years, the expense for this technician can be included in the purchase price of all the machine tools. If an OEM is anxious to sell multiple tools, this technician will be supplied at a reduced rate. If your company purchases multiple tools from the same OEM, it is reasonable to negotiate with the OEM to provide the technician to assist with the installation and maintenance of the tool until the last machine is out of warranty.

7.8   One-Year Meeting after Final Acceptance— Lessons Learned

Schedule a meeting with the OEM and your team 1 year after the installation of the machine tool. At this time, voice your concerns on items the OEM can help correct at the end of the warranty period. Schedule this meeting at least 6 months in advance to allow the OEM and your engineering, maintenance and facilities, and operational departments time to detail every issue associated with the tool. There are usually few, if any, issues to discuss with the OEM at this point, but it is a courtesy to all departments to have the opportunity to discuss any lingering issues or new concerns with the new machine tool.

7.9   Documentation of the Machine Tool

You can never have too much documentation of the machine tool. It is inconvenient and inefficient to consult with the OEM each time a question arises. Here is a list of documentation the OEM should supply with the tool:

1. *Hard-copy documentation requirement:* The OEM should provide at least three hard copies of all items.

2. *Operations manual for the machine tool:* The OEM should detail all operator control panel buttons and their operation. The manual should detail the operation of any special screens on the CNC control and operation.

3. *Maintenance manual for the machine tool:* The most important section of the maintenance manual is the OEM's list of CNC alarms, which is generated from the OEM's functions.

4. *Mechanical drawings of the machine tool:* Get as many detailed drawings as possible. Request that the OEM provide more than just the mechanical assembly drawings of the machine tool.

5. Electrical drawings of the machine tool

6. *Drawings of all pieces of equipment that are consumable:* This includes special tool holders, gibs, centers, retention knobs, tool tapers, and others.

7. *Preventive maintenance manuals:* Secure the recommended preventive maintenance manuals for the machine tool.

8. CNC programming manuals from the CNC manufacturer

9. Foundation drawings and general arrangement drawings

10. Chip conveyor drawings from the OEM or from the chip conveyor manufacturer

11. Mechanical drawings of all alignment tools used for aligning and calibrating the machine tool

12. *Detailed information for all electrical and mechanical components:* For example, know who supplies the necessary fuses and circuit breakers.

7.10   Summary

In summary, the start-up phase includes the entire machining, test parts, and the continuous running tests. It checks the functions of the main control panel and operator control panel. It allows for gathering as much information from the OEM as possible. Get the contact information of the OEM's service manager, including e-mail, work number, and cell phone number. Get the machine tool supplier's programming support hot-line phone and e-mail contact information. Get the night and weekend support contact information from the machine tool supplier in case the machine tool goes down on a weekend. Pass this information onto the maintenance and facilities, engineering, and operational departments.

CHAPTER **8**

Maintenance Phase

8.1　Why Perform Maintenance?

As astronomical an expense as it is to do everything necessary to purchase the right machine tool and install it correctly, the spending doesn't end there—not by a long shot! After your company purchases and installs an expensive machine tool, maintaining it properly should not cost a small fortune. It's maddening to witness, as I often have, a company spend millions of dollars and thousands of worker-hours getting the new machine tool bought and up and running, and then give no thought as to what is required to keep it producing money for the next 20 years. I suppose that, after the buyer spends such a huge sum of capital funding on a new machine tool, foundation, and the countless other necessities, the very thought that endless spending will be required to keep the machine tool working at maximum productivity is enough to make any executive want to crawl under the desk and cry! Who can blame that executive? But how easy it is to forget that the company spent all this money so that it could recoup its investment many times over during the life of the machine. The shorter the machine's life, the less money it will produce, to say nothing of the cost of replacing it.

8.2　Decrease Machining Downtime by Implementing a Daily Operational and Preventive Maintenance Program

By implementing a daily operational and monthly, semiannual, and annual preventive maintenance program, there will be a decrease in the number of repairs, thus decreasing overall maintenance costs and machining downtime. The preventive maintenance program should not be subject to the everyday bureaucratic policies of a typical manufacturing management. The process should be simple and unhindered. The following is a very basic preventive maintenance schedule:

- Daily operator checklists
- Monthly maintenance checklists
- Semiannual maintenance checklists
- Annual maintenance checklists

The daily operational checklists should be turned in to the operations supervisor at the end of every shift. The supervisor should quickly check the list and review each item. The maintenance supervisor or maintenance engineer should review and confirm that each item was completed and signed off by the mechanic, electrician, or technician. Once this step is completed, the supervisor should then sign off on it herself or himself. This should be the procedure for every preventive maintenance checklist.

The maintenance mechanic, electrician, or technician should detail any issues that arise during the completion of the checklist. There should be a place on the form to document issues not provided on the checklist form. This should include any spare parts needed; whether they are obtained by in-house reserves or have to be purchased, they should definitely be recorded on the list. In short, the checklist must provide a detailed history of everything the machine tool has required and what was done to address it, for the period the checklist is reporting on.

All the checklists listed above are equally important. The daily one will go a long way in determining how the others will read if it ensures that new problems are addressed immediately. There should also be an annual review of both the operational and preventive maintenance checklists to make sure that your company is continually reviewing and improving the preventive maintenance process.

8.3 Increase Overall Productivity of the Machine Tool

Most large machine shops use giant machine tools costing millions of dollars. Nearly all run multiple shifts of machinists.

The machine tools are running 24 h/day, 7 days/wk. It is imperative that proper operational and maintenance countermeasures be employed to keep these machines running smoothly. This will definitely increase the life of any machine and reduce significantly the overall cost of manufacturing.

The top priority in every company's operations is that the spindle of the HBM or lathe is ready to turn anytime and every time it is called upon to do so. Daily preventive maintenance is simple and comparatively cheap. For example, all machine tools have a filtration system for the hydraulics. There is always some filter or strainer in line to clean the oil while the machine tool is running. The machine tool operator should always keep spare filters in the tool crib for each machine tool.

8.4 Higher Accuracy and Quality of Parts Produced by Checking Your Machine Tool's Alignments and Laser Calibration at least Once per Year

A well-serviced, -aligned, and -calibrated machine tool will produce accurate and high-quality workpieces. It is critically important to check the mechanical alignment of the machine tool's bed, column, ram, and spindle (quill or bar) if you have an HBM. On a horizontal lathe, the mechanical alignment of the cross slide, bed, and headstock should be inspected on an annual basis. Spindles on lathes and HBMs should be checked for run-out and out-of-round conditions.

The calibration of the spindle and headstock is just as important as that of any other part of the machine tool. By using electronic levels, granitite parallels, and leveling wedges, you should check the machine's mechanical accuracies to earth, based on the OEM's specifications.

Check the machine tool's linear accuracies with a laser. The laser will check the incremental linear straightness of the machine in single or multiple directions. Most of the newer CNCs have the ability to compensate for the linear inaccuracies of the machine tool in one or multiple axes. The newer CNCs also have temperature compensation for large temperature gradients that may occur in your facility or surroundings.

Other checks that can be performed are as follows:

1. Machine tool CNC positioning accuracy for all the linear and rotary axes

2. Machine tool axis straightness (bed, column, headstock, etc.)

3. Machine tool axis angular errors (bed, column, headstock, etc.)

8.5   Daily Operations Checklist

The daily operations checklist is something the machinist or operator should complete on every shift. These items should be checked daily. The operations department is responsible for checking the oil and lubrication of the machine tool daily. Every tank on the machine tool should be clearly labeled for the exact type of oil needed for that tank, and that should be checked daily. Figure 8.1 is a list of other items that should be checked daily by operations:

Item to Be Checked	Sign-off, Shift, and Date
1. Check for oil and hydraulic leaks around the entire machine tool.	
2. Check oil hydraulic tanks for proper levels.	
3. Check coolant tanks for proper levels.	
4. Remove and clean all metal chips from around the machine tool.	
5. Check the condition of the metal chip bin.	
6. Remove and clean the metal chips from the telescopic way covers of the machine tool.	
7. Check the incoming air pressure for the machine tool.	
8. Check that all tools and tool holders have been properly placed in the correct location.	
9. Check the condition of the ways for each axis for damaged or metal chips.	
10. Detail any issues from the current shift in the Notes section.	
Is the machine tool in operation at shift change? If yes, do not check the daily operational items. If no, you must check the daily operational items.	YES or NO Circle one
Notes:	

FIGURE 8.1   Daily operational checklist and sign-off (per shift).

8.6   Monthly Maintenance Checklist

Figure 8.2 is a standard list of monthly maintenance checks that the maintenance and facilities group should be making. This is an example of the items that need to be checked monthly on your company's requirements and capital funding. It is extremely important that all oil, lube, and air filters be documented on the machine tool, and each filter should be stocked in the tool crib.

8.7   Semiannual Maintenance Checklist

The semiannual maintenance checklist should be checked every 6 months. Figure 8.3 shows a standard list of monthly maintenance checks that the maintenance and facilities group should be checking. This is example of the items that needs to be checked semi-annual basis on your company's requirements and capital funding.

8.8   Annual Maintenance Checklist

The main tasks to be performed during the annual maintenance checklist are the alignment and laser calibration of the machine tool. The work in the annual maintenance checklist should be done yearly and could last up to 2 weeks for a large machine tool. It is important that the machine tool's alignment and laser calibration data be checked yearly to make sure that the foundation hasn't settled or that the crane operator hasn't dropped something on the machine tool bed or rotary table.

This is also the time when you plan to repair or replace damaged parts that required you and your company to purchase spare parts that had a long lead time. Figure 8.4 shows a standard list of annual maintenance checks that the maintenance and facilities group should be performing. This is an example of the items that need to be checked monthly based on your company's requirements and capital funding. Since all the items on daily, monthly, semiannual, and annual checklists are examples, they can be duplicated as many times on each of the checklists as needed.

Item to Be Checked—Mechanical	Sign-off, Shift, and Date
1. Check for oil and hydraulic leaks around the entire machine tool.	
2. Check oil hydraulic tanks for proper levels.	
3. Check coolant tanks for proper levels.	
4. Remove and clean all metal chips from around the machine tool.	
5. Check the operation of the hardware and software limits. Document the software and hardware limits in the Notes.	
6. Remove and clean off metal chips from the telescopic way covers of the machine tool.	
7. Check the oil level in all the gearboxes.	
8. Check the belts or gears of each motor on the machine tool. This is an item in which you should detail each motor on the machine tool.	
9. Replace all oil and lube filters on the machine tool.	
10. Replace all the air filters on the machine tool.	
11. Check and replace as needed all pressure gauges on the machine tool. Also document the pressure at each location in the Notes.	
12. Check the filters for spindle motor, axis motors, hydraulic motors, and air conditioning units	
Items to Be Checked—Electrical	**Sign-off, Shift, and Date**
1. Check the overall travel limits of the machine tool. Make sure that the machine tool alarms on the software limits before the hardware limit.	
2. Inspect all power tracks on the machine tool.	
3. Turn off power to the machine tool, and visually check inside the main electrical cabinet for damaged electrical wire or cables.	
4. Check all encoder or tachometer connections.	
5. Check indicator lights and buttons on the operator panel.	
6. Check the operation and condition of each motor and air conditioner on the main control panel.	
7. Turn off the power to the main control panel, and tighten all mechanical connections.	
Is the machine tool in operation at shift change? If yes, do not check the daily operational items. If no, you must check the daily operational items.	YES or NO Circle one
Notes:	

Figure 8.2 Monthly maintenance checklist and sign-off.

Item to Be Checked—Mechanical	Sign-off, Shift, and Date
1. Check for oil and hydraulic leaks around the entire machine tool.	
2. Check oil hydraulic tanks for proper levels.	
3. Check coolant tanks for proper levels.	
4. Remove and clean all metal chips from around the machine tool.	
5. Check the clamping pressure of the drawbar on the spindle.	
6. Remove and clean off metal chips from the telescopic way covers of the machine tool.	
7. Check the oil level in all the gearboxes.	
8. Check the belts or gears of each motor on the machine tool. This is an item for which you should detail each motor on the machine tool.	
9. Replace the oil, lube, and air filters on the machine tool.	
10. Check all way wipers on each axis (both sides).	
11. Check the ac or dc axis motor drive belts on each axis and spindle motor.	
12. Check all pump couplings.	

Items to Be Checked—Electrical	Sign-off, Shift, and Date
1. Check all ac or dc motor couplings.	
2. Inspect all power tracks on the machine tool.	
3. Grease all motor bearings for each axis.	
4. Check all encoder or tachometer connections and couplings.	
5. Check indicator lights and buttons on the operator panel.	
6. Replace the air filters on each motor and air conditioners on the main control panel.	
7. Check the brushes and commutators for all dc motors.	
8. Clean and inspect all linear scales and reader heads for each axis.	
9. Clean all air intakes on all ac or dc motors.	
10. Clean all air intakes on the air conditioning systems on the main electrical control cabinets.	
11. Check the hard overtravel switches for each axis and on both sides (positive and negative direction).	

Is the machine tool in operation at shift change?	YES or NO
If yes, do not check the daily operational items.	Circle one
If no, you must check the daily operational items.	
Notes:	

FIGURE **8.3** Semiannual maintenance checklist and sign-off.

Item to be Checked—Mechanical	Sign-off, Shift, and Date
1. Perform a complete alignment check of the machine tool per the machine tool supplier's standards.	
2. Check oil hydraulic tanks for proper levels.	
3. Check coolant tanks for proper levels.	
4. Remove and clean all metal chips from around the machine tool.	
5. Check the condition of the headstock and tailstock for each chip conveyor associated with the machine tool.	
6. Remove the telescopic way covers of the machine tool from the machine tool, and visually inspect the bed of the machine tool. Perform this operation for each axis that has a telescopic way cover	
7. Check the oil level in all the gearboxes.	
8. Check the belts or gears of each motor on the machine tool. This is an item for which you should detail each motor on the machine tool.	
9. Replace all oil and lube filters on the machine tool.	
10. Replace all the air filters on the machine tool.	
11. Check and replace as needed all pressure gauges on the machine tool. Also document the pressure at each location in the Notes.	
12. After completion of the mechanical section of the annual preventative maintenance (PM) turn on the machine tool to verify that the machine tool properly operates and references.	
Items to Be Checked—Electrical	**Sign-off, Shift, and Date**
1. Perform a complete laser calibration of the machine tool per the machine tool supplier's standard specifications.	
2. Physically tighten all electrical connections in the main electrical control cabinet.	
3. Turn off power to the machine tool, and visually check inside the main electrical cabinet for damaged electrical wire or cables	
4. Remove the end of the linear scales, and visually inspect the scale housing.	
5. Check of indicator lights and buttons on the operator panel.	
6. Remove the electrical cover of the main spindle motor, and visually inspect the main electrical connections between the spindle motor and main control panel.	
7. After completion of the electrical items, turn on the machine tool and make sure that the machine tool properly operates and references.	
Is the machine tool in operation at shift change? If yes, do not check the daily operational items. If no, you must check the daily operational items.	YES or NO Circle one
Notes:	

FIGURE 8.4   Annual maintenance checklist and sign-off.

8.9   Standard Spare Parts List

Now that your company has spent millions of dollars on a new machine tool, expect to spend another 5 to 10 percent on the most commonly needed spare parts to have in stock. Either purchase them from the machine tool supplier, or have the purchasing department procure them after the installation of the machine tool is completed. Of these two options, the less expensive one is to have the purchasing department research and purchase them from various original equipment suppliers. That's where the machine tool supplier will get them too; then the supplier will mark up the price and pass it all on to your company.

That said, it is sometimes a good idea to purchase the first set of spare parts from the machine tool supplier. This serves two purposes. First, it provides your purchasing department with a record of which spare parts should be kept in reserve on the premises. It also helps purchasing determine where and where not to buy spare parts.

Here is a list of standard parts you should stock:

- Oil and lubrication element filters
- Air filters
- Oil
- Lubrication
- Grease
- Fuses (at least 12 for three-phase circuits and 10 for single-phase circuits)
- Proximity switches
- Tool changer holder
- Tool probe
- Tool holders
- Retention knobs for the tool holders
- Spare Belleville washers for the spindle clamping device or spare drawn bar of the spindle
- Spare keys for the spindle face
- Spare lubrication motor

- Spare scavenger pump for the headstock or bed hydrostatics
- At least two or three spare handheld units and manual pendants
- Single-phase circuit breakers
- Three-phase circuit breakers
- Gibs

8.10   Long-Term Spare Parts List

As with the standard parts list above, you can obtain long-term spare parts from the machine tool supplier or have the purchasing people procure them. Again, the less expensive route is to have the purchasing department purchase them from various original equipment suppliers. That's where the machine tool supplier will get them; then the supplier will mark up the price and pass it on to your company. Here is a long-term spare parts list:

- Bearings for the headstock
- Spare axis feed motor for each axis on the machine tool
- Spare glass or tape scale for the each axis
- Spare reader head for the glass or tape scale for each axis
- Spare axis drive for each axis
- Spare encoder for the ram and/or spindle of the headstock
- Spare spindle drive
- Spare spindle motor
- Spare Turcite for the bearing and boxways
- Bearings for the tailstock
- Spare CNC monitor
- Spare CNC membrane or mechanical keyboard
- At least two or three spare spindle and feed rate override switches
- Spare PLC inputs card (one of each type)
- Spare PLC output card (one of each type)

The long-term spare parts list is at least 5 times more expensive than the standard spare parts list. These parts not only are costly but also usually take months, not weeks, to manufacture and deliver. Some needed parts may be out of production for weeks and months, not hours and days. If they are not on hand, a multimillion-dollar machine tool could be out of service for a very long time, hindering or even crippling your company's production operation.

8.11   Summary

In summary, implementing daily operational and preventive maintenance programs will make or break the long-term operational capabilities of new machine tools. It could also thwart the broader long-term goals of the company. A nonfunctioning machine tool or, worse yet, machine tools, will always result in unmet goals such as targets for sales, safety, and hours worked; this is not to mention the unhappy machinists who have to work long hours to make up for the downtime due to lack of machine maintenance, and so much more.

The importance of having a good operational and preventive maintenance program, having on hand the spare parts needed to keep the machine tool running, and having a management team who understands just how crucial maintaining your equipment is, cannot be overstated. Without all these, keeping the machine tools running and the company competitive will always be an uphill battle.

Final Summary of the Book

The goal of this book is to provide a way for individuals in manufacturing facilities to develop a process for specification, purchase, installation, and maintenance of the machine tools needed for the future. Hopefully, you can take some ideas from this book and use them in a future machine tool purchase. Even if it is just one or two ideas, it could save you and your company precious time and funding, which could then be used for some other piece of equipment or tooling.

It has always been my goal, wherever I have worked, to increase productivity, reduce the downtime of the equipment, and save the company money in any way possible. By writing this book and detailing the various schedules, checklists, specifications, and concepts, I hope to help you streamline the process for the purchase and installation of the new machine tool. It doesn't matter whom you work for or what country you live in. I want to save you and your company time and money.

I truly believe that in every manufacturing industry around the world there are good stewards of continuous improvement. I think that this book helps start a process of budgeting, specifying, purchasing, installing, starting up, and maintaining the machine tool for many years to come. See below the checklist of items for the budgeting, specification, and installation of the foundation and machine tool, as well as preparation, start-up, and maintenance of a new machine tool. These are the standard basic items that you should focus on during the overall process of buying a machine tool.

Checklist for the Final Summary Chapter	
Budget Phase	**List of Items to Perform during the Budget Phase**
	Attend trade shows—IMTS, EMO, and South Tech.
	Develop a return on investment.
	Develop a payback.
	Develop a breakeven cost.
	Contact potential machine tool suppliers.
	Draw up the first-draft specifications.
	Gather "ballpark" quotes.
	Deliver a cost estimate for the new machine tool and present it to upper management for approval.
Procurement Phase	**List of Items to Perform during the Specification Phase**
	Develop the final machine tool specification.
	Ask the machine tool manufacturers for real quotes based on the machine tool specification.
	Quote a minimum of 5 machine tool suppliers. In my opinion, you should involve at least 10 machine tool suppliers if you have enough time or if the project or machine tool is extremely large.
	Compare all the machine tool suppliers' quotes side by side in spreadsheet form.
	Pick the top two machine tool suppliers and negotiate the final costs. "Best and final" quotes will be asked of the final two machine tool suppliers at this time.
	Supply the machine tool supplier with the amount of training needed and training lists (operations, maintenance, and applications).
	Create the machine tool purchase order.
Layout Phase	**List of Items to Perform during the Layout Phase**
	Lay out the machine tool in at least two or three different orientations in the area where you plan to locate the machine tool. These layout drawings need to be in both the plan and elevation views to show any trouble conditions between the new machine tool and the building or other machine tools.
	Maximize the floor space by moving the main electrical panel and auxiliary equipment in the column line of the building. This might require the buyer to add more trenches and ask the machine tool supplier to separate the new electrical panels if needed.

	Checklist for the Final Summary Chapter
	Review the general Notes section of the machine tool supplier's general arrangement and foundation drawings. Make sure they get transferred to the plant layout drawings and the new foundation drawings that will be developed to build the new foundation.
	What is the maximum weight of the machine tool? What is the maximum weight of the largest and heaviest piece of the machine tool? What is the heaviest workpiece you can machine on the machine tool? Transfer all this information to the designer of the new foundation.
	Develop a soil-bearing specification for the facility in which you work.
	Minimum requirements for locating a machine tool in a factory or manufacturing facility.
	Develop the overall schedule for the layout and foundation phases.
	Plan to remove the old machine tool, old foundation, and old utilities in the location where the new machine tool is to be located.
Foundation Phase	**List of Items to Perform during the Specification Phase**
	Who will lead the foundation specification, purchase, and installation?
	Complete the foundation installation schedule.
	Double check the trench design, sump design, chip conveyor trenches, and electrical panel locations versus the foundation design of the machine tool. These items must match with the layout of the machine tool in which you are purchasing.
	Develop a foundation specification for the new machine tool.
	Get at least five design reviews for the foundation for the new machine tool.
	Bid the new foundation design and place the purchase order.
	Install the dust walls and limit access to the build of the foundation.
	Prejob safety brief with the general contractor before and monthly with the builder of the machine tool foundation.
	Implement a Concrete cylinder test for the concrete that is used in the pour of the machine tool foundation
	Don't forget that concrete has a 28 day period of time to reach 99% of its compressive strength.

(Continued)

Checklist for the Final Summary Chapter	
Installation Phase	**List of Items to Perform during the Specification Phase**
	Develop a machine tool installation specification, and ask for a bid from three to five potential machine tool installation companies.
	Develop a schedule for the installation of the machine tool.
	Develop work hour limits for the machine tool installation company.
	Place the purchase order for the installation of the machine tool at the buyer's facility.
	Do the epoxy painting of the machine tool foundation.
	Purchase the oils and coolants for the new machine tool.
	Purchase and install the machine tool metal chip bin.
	Purchase the grout and grout forms.
	Design and purchase the safety rails and ladders around the new machine tool.
Preparation Phase	**List of Items to Perform during the Specification Phase**
	Chip conveyor requirements: Is the chip conveyor being supplied by the machine tool supplier, or will the buyer of the machine tool supply the chip conveyor?
	Chip conveyor baffles or deflector requirements: Does the new machine tool need deflectors or baffles?
	Main power source for the new machine tool: What is the needed three-phase power, 480 or 230 VAC?
	Secondary power source for the new machine tool: Does the new machine tool need 110-VAC single-phase power?
	Is an isolation transformer needed between the new machine tool and the main power source of the facility?
	Do you need to purchase a main breaker which supplies power to the new machine tool?
	Determine the compressed air and water requirements for the new machine tool.
	Does your new machine tool require so type of oil skimmer? The types that you can use are Disk, belt, or tramp oil skimmer systems
	Will a coolant enclosure or wall be needed around the machine tool?

Checklist for the Final Summary Chapter
If your machine tool foundation has a pit under the machine tool it will require lighting in the pits of the machine tool foundation.
Does the machine tool need a direct numerical control (DNC) software package for uploading and downloading for the part programs?
Create a procedure to upload and download the part program using the DNC software.
Install the "structure steel" around the machine tool, e.g., checkered plate, diamond plate, and bar grating.
Perform alignment and laser alignments on the new machine tool.
Get the final machine tool acceptance.
Perform the first machining test.
Develop a part program and test part program.
Connect and test the teleservice for the machine tool supplier.
Detail the beginning of the warranty period
What type of conduit is to be used, RMC, IMC, EMT, or PVC?
Does the machine tool supplier expect the machine tool buyer to supply the power track or wireway around the new machine tool foundation?
If the new machine tool has a head attachment house, test each of the new head house pickup and drop-off locations at least 10 times for each head attachment.
Understand the sag compensation and temperature compensation of the new machine tool.
Check the hard and soft overtravel limits of the new machine tool.
Perform the basic reference of the new machine tool.
Detail the procedure to start up the machine tool from an off condition.
Has a lock out and tag out procedure for the new machine tool been created?
Back up all the CNC data of the new machine tool after completion of the machine tool installation phase.
Specify the procedure to view and access the PLC ladder logic of the new machine tool.

(Continued)

Checklist for the Final Summary Chapter	
Start-up Phase	**List of Items to Perform during the Specification Phase**
	Have the machining test and test part been completed, and did it pass the acceptance test?
	Perform additional acceptance tests.
	Do the final machining tests.
	Will the machine tool supplier have on-site support during the first year of operation?
	Call the one-year meeting after installation of the machine tool. This is a "lessons learned" meeting with the machine tool supplier and the operational department where the machine tool is located.
	Has the final documentation of the new machine tool been received and accounted for?
Maintenance Phase	**List of Items to Perform during the Specification Phase**
	Implement a daily operation checklist.
	Implement a monthly maintenance checklist.
	Implement a semiannual maintenance checklist.
	Implement an annual maintenance checklist.
	Perform a machine tool alignment and laser accuracy check every year on the new machine tool.
	Develop a standard spare parts list.
	Develop a long-term spare parts list.

Index

Note: Page numbers followed by *f* denote figures; page numbers followed by *t* denote tables.